Keio University International Symposia for Life Sciences and Medicine Vol. 13
H. Ishii, M. Suematsu, K. Tanishita, H. Suzuki (Eds.)
Organ Microcirculation
A Gateway to Diagnostic and Therapeutic Interventions

H. Ishii, M. Suematsu,
K. Tanishita, H. Suzuki (Eds.)

Organ Microcirculation

A Gateway to Diagnostic and Therapeutic Interventions

With 93 Figures, Including 2 in Color

Springer

Hiromasa Ishii, M.D., Ph.D.
Professor Emeritus, Department of Internal Medicine
School of Medicine, Keio University
35 Shinanomachi, Shinjuku-ku, Tokyo 160-8582, Japan

Makoto Suematsu, M.D., Ph.D.
Professor, Department of Biochemistry and Integrated Medical Biology
School of Medicine, Keio University
35 Shinanomachi, Shinjuku-ku, Tokyo 160-8582, Japan

Kazuo Tanishita, Ph.D.
Professor, Department of System Design Engineering
Faculty of Science and Technology, Keio University
3-14-1 Hiyoshi, Kohoku-ku, Yokohama, Kanagawa 223-8522, Japan

Hidekazu Suzuki, M.D., Ph.D.
Assistant Professor, Department of Internal Medicine
School of Medicine, Keio University
35 Shinanomachi, Shinjuku-ku, Tokyo 160-8582, Japan

Library of Congress Control Number: 2004113090

ISBN 4-431-22135-2 Springer-Verlag Tokyo Berlin Heidelberg New York

Springer is a part of Springer Science+Business Media
springeronline.com

Printed in Japan

Typesetting: SNP Best-set Typesetter Ltd., Hong Kong
Printing and binding: Shinano Inc., Japan

Printed on acid-free paper

Foreword

This volume of the *Keio University International Symposia for Life Sciences and Medicine* contains the proceedings of the 14th symposium held under the sponsorship of the Keio University Medical Science Fund. The fund was established by the generous donation of the late Dr. Mitsunada Sakaguchi. The Keio University International Symposia for Life Sciences and Medicine constitute one of the core activities sponsored by the fund, of which the objective is to contribute to the international community by developing human resources, promoting scientific knowledge, and encouraging mutual exchange. Each year, the Committee of the International Symposia for Life Sciences and Medicine selects the most significant symposium topics from applications received from the Keio medical community. The publication of the proceedings is intended to publicize and distribute the information arising from the lively discussions of the most exciting and current issues presented during the symposium. On behalf of the Committee, I am most grateful to the late Dr. Sakaguchi, who made the series of the symposia possible. We are also grateful to the prominent speakers for their contributions to this volume. In addition, we would like to acknowledge the efficient organizational work performed by the members of the program committee and the staff of the fund.

Naoki Aikawa, M.D., D.M.Sc., F.A.C.S.
Chairman
Committee of the International Symposia
for Life Sciences and Medicine

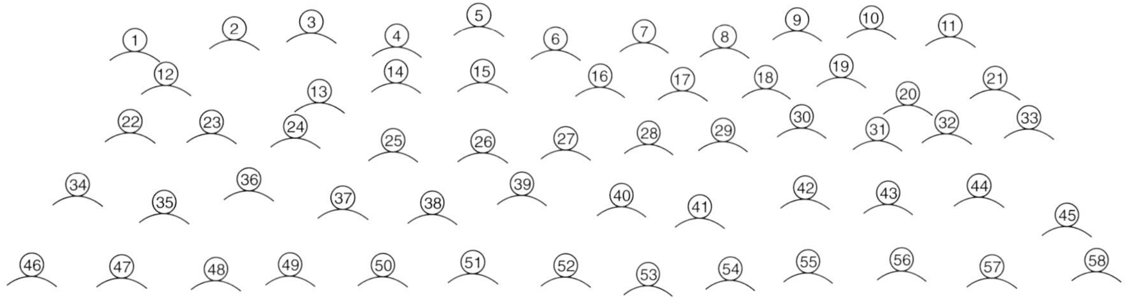

The 14th Keio University International Symposium for Life Sciences and Medicine

1. Mayumi Kajimura 2. Motohide Shimazu 3. Yoshiyuki Takei 4. Takeharu Shigematsu 5. Dai Fukumura 6. Mikiji Mori
7. Satoshi Terao 8. * 9. Mami Ishikawa 10. * 11. Satoshi Takeuchi 12. Tomihiro Katayama 13. Masayuki Suzuki
14. Yoshikazu Tsuzuki 15. Yoshinori Horie 16. Tetsuro Yamamoto 17. Kazuo Tanishita 18. Takashi Osada 19. Hidenori Iwasaki
20. Misa Mizumori 21. Manabu Nakashita 22. Shinji Takeoka 23. Sumito Mizuguchi 24. Hiroyuki Katagiri 25. Masataka Majima
26. Akira Fujimura 27. Souichiro Miura 28. Takashi Joh 29. Masahiko Nakamura 30. Ric Anarsons 31. Arifumi Iwamaru
32. Hiroshi Nagata 33. Hidekazu Suzuki 34. Makoto Suematsu 35. Makoto Katori 36. Haruyuki Minamitani 37. Chiyoji Ohkubo
38. Makoto Ito 39. Minoru Tomita 40. Norio Ohshima 41. Chikara Ooshio 42. Tetsuo Morishita 43. Norio Tanahashi
44. Nobuhiro Sato 45. Rina Takamiya 46. Fumihiko Kajiya 47. C. Charles Michel 48. Paul Kubes 49. Michael D. Menger
50. Geert W. Schmid-Schönbein 51. Robert S. McCuskey 52. Hiromasa Ishii 53. D. Neil Granger 54. Andrzej S. Tarnawski
55. Mark G. Clemens 56. Rodger P. McEver 57. Jonathan D. Kaunitz 58. Hidekazu Tsukamoto

* The editors regret that the names were unavailable at the time of publication.

Preface

The 14th Symposium for Life Sciences and Medicine of the Keio University Medical Science Fund was held December 11–13, 2003. The symposium focused on the recent progress achieved in research in the field of organ microcirculation, with the main theme being "A Gateway to Diagnostic and Therapeutic Interventions". The aim of the symposium was to present and exchange the most recent information and opinions by scientists in Japan and other countries. The target was the confluence of medical engineering technology using nanotechnology, where there recently have been striking advances, and the latest information in molecular biology. This symposium was held in collaboration with the Faculty of Science and Technology of Keio University to combine applied biomedical technology with interventional medicine that targets organ microcirculation.

The symposium was a success because of the large body of participants—a total of 100, including 16 distinguished scientists from abroad (13 from the United States, 1 each from United Kingdom, Germany, and Canada). Twenty-five papers were presented orally and 18 poster presentations were made. Six poster presentations were selected for Excellent Investigators Awards.

All the participants joined in the discussions about current topics in diverse aspects of organ microcirculation research. This volume presents all those topics and the essence of the discussions during the symposium.

I strongly believe that this symposium has contributed significantly to the advancement of microcirculation research worldwide. I would like to thank all the contributors for sharing their knowledge and expertise in assembling these proceedings. The symposium would not have been possible without the support of the Keio University Medical Science Fund established by the generous donation of the late Dr. Mitsunada Sakaguchi. I express my deep appreciation to the founder, and I am grateful for all the efforts of the staff members of the fund. Finally, I thank the staff of Springer-Verlag, Tokyo, for their excellent work in publishing this volume.

Hiromasa Ishii, M.D., Ph.D.
Chairman
The 14th Keio University International Symposium for
Life Sciences and Medicine
Professor Emeritus
Keio University

Contents

Part 2 Gastroduodenal Microcirculation and Disease

Part 4 Cell Adhesion and Traffics in Microcirculation

List of Contributors

Opening Remarks

Professor Yuichiro Anzai
President, Keio University
Chairman, Keio University Medical Science Fund

On behalf of Keio University, it is a great pleasure to welcome all of you to the 14th Keio University International Symposium for Life Sciences and Medicine. I am particularly grateful to the distinguished medical scientists, speakers, and others who are participating in this meeting.

Although remarkable progress and breakthroughs are reported almost every day in the research field of microcirculation, it is also true that there are still some patients who suffer from various kinds of organ failure due to microvascular disturbances. In this regard, I believe that it is an opportune moment to hold a symposium at Keio University related to the subject. All speakers kindly accepted our invitation to contribute to this symposium on organ microcirculation, and I feel certain that this unique meeting will prove both exciting and successful.

Keio was founded in 1858 by Yukichi Fukuzawa and is the oldest institution for modern education and research in the country. Fukuzawa was a pioneer in the modernization of Japan. International exchanges such as this symposium have been one of the most important academic and social missions of Keio University since its foundation.

In 1994 and also in 1999, Dr. Mitsunada Sakaguchi, an alumnus of the class of 1940 of our medical school, donated a significant amount of money to the university. He expressed the wish that his gift be used to encourage research in life sciences and medicine and to promote worldwide advancements in life sciences. We agreed with his proposal and thus launched the Keio University Medical Science Fund in April 1995. The International Symposia for Life Sciences and Medicine have been organized as one of several projects supported by the fund.

It is thus more than a pleasure—indeed, it is an honor—for me to be able to meet the distinguished medical researchers and clinicians from world-renowned institutions who have kindly gathered here, and to share in frank and valuable exchanges of views. I am also grateful for the efforts made by the organizing committee, chaired by Dr. Hiromasa Ishii, who has devoted himself to ensuring that this symposium is an auspicious and enjoyable event. I do hope that the meeting will prove a truly fruitful and productive one for you all.

I wish everyone gathered here further success in your research and clinical work. Thank you very much.

Part 1
Nanobiotechnology for Developing Microvascular Intervention

Importance of the Villus Microcirculation for Intestinal Absorption of Glucose

C. Charles Michel[1] and John R. Pappenheimer[2]

Key words. Villus microcirculation, Intestinal absorption, Glucose, Permeability, Epithelia

Introduction

The microcirculation is often regarded as the junior partner in the spectacular transport processes that occur in the gastrointestinal tract. While its importance in the delivery of oxygen is acknowledged, its role in the delivery of solutes other than oxygen to sites of secretion and in their clearance from sites of absorption is often considered briefly or regarded as obvious and dismissed without further discussion. We have recently conducted an analysis of intestinal glucose absorption and our results suggest that this neglect of the microcirculation is misplaced [1]. Increases in blood flow through the villus microcirculation in proportion to increases of glucose absorption accompanied by similar increases in the product of permeability and surface area of the exchange vessels, appear to be essential for high rates of glucose uptake. In this paper we summarize our conclusions and outline the basis of the analysis we have used to reach them.

Principles of the Analysis

We have considered epithelial and microvascular transport of glucose as two processes in series. We have assumed that a steady state is rapidly established between glucose entry through the brush border of the epithelial cells and the

[1] Division of Biomedical Sciences, Faculty of Medicine, Imperial College, Exhibition Road, London SW7 2AZ, UK
[2] Department of Biology, Harvard University, Cambridge, MA 02138, USA

transport of glucose away from the mucosal region by the microcirculation of the intestinal villi. To estimate glucose concentration at various points between the brush border and the blood flowing out of the villus capillaries, we have taken a well-defined model for the glucose pathways. While some of the details of the epithelial section of the pathway are not universally accepted, changing these details does not affect our conclusions.

Our basic model is shown in Fig. 1. Diagrammatically, it portrays two epithelial cells from the jejunal section of the intestine where glucose is absorbed. The brush border of the apical membranes and the villus capillaries adjacent to the basement membranes are indicated and the basal three-quarters of the cells are truncated and separated by the lateral intercellular space (LIS).

Glucose is taken up through the brush border of the epithelial cells with Na^+ on the SGLT-1 transporter where it is concentrated in the apical regions of the cells. The Na^+ is pumped out of the cells into the lateral intercellular spaces (LIS) beneath the tight junctions on Na^+-K^+ ATPase in parallel with the glucose, which is carried out of cells on the Glut-2 transporter by facilitated

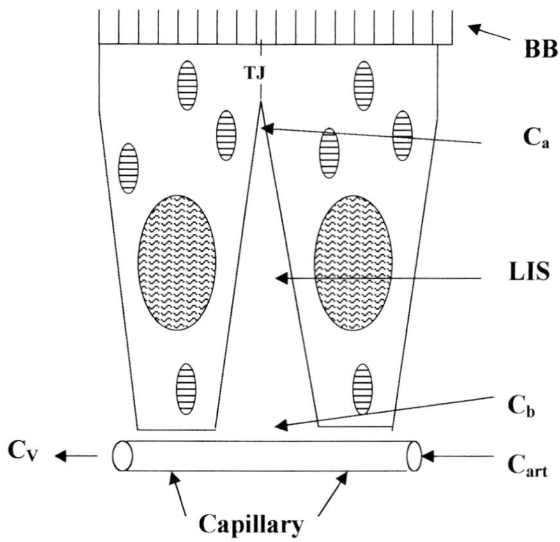

FIG. 1. Diagram of two jejunal epithelial cells and their relation to the villus capillaries. The brush border (*BB*) is indicated in the apical membranes of the cells and the upper quarter of the cells are closely opposed and joined by tight junctions (*TJ*). The basal three quarters of the cells are roughly conical and separated by large intercellular spaces (*LIS*). Other abbreviations are: C_a, glucose concentration at top of LIS just beneath TJ; C_b, glucose concentration in LIS at level of epithelial basement membrane; C_{art}, C_V, arterial, venous glucose concentrations

diffusion. The Glut-2 transporters are located immediately beneath the tight junction and the glucose molecules arriving here via Glut-2 are joined by more glucose molecules that have been carried by solvent drag through the junction. Glucose then passes by diffusion and convection down the LIS to the epithelial basement membrane adjacent to the villus capillaries. Glucose diffuses into the capillaries and is carried away by the blood flow.

In the first and most important part of our analysis, we have used measured rates of glucose absorption in conscious rats and human subjects to calculate mean glucose concentrations in the LIS immediately beneath the tight junction (C_a), at the epithelial basement membrane just outside the villus capillaries (C_b), in the villus capillary blood (C_m), and in the blood leaving these capillaries (C_V).

In the second part of our analysis we have estimated the fraction of the glucose passing through the apical regions of the epithelial cells and using the properties of Glut-2 transporters, we have made tentative estimates of the intracellular glucose concentration (C_{cell}).

Equations for Calculation of Glucose Concentration Along the Transport Pathway

Our fundamental assumption is that the rate of glucose entry at the apical surface of the epithelium (J_S) is the same as the glucose flux away from the villi in the blood. If F is the villus blood flow, C_{art} is the arterial concentration of glucose, C_V the venous glucose concentration, then by the Fick principle:

$$C_V = C_{art} + J_S/F \tag{1}$$

If C_{art}, C_V, F and the product of permeability and surface area of the villus capillaries, PS, are known, the concentration of glucose in the interstitial fluid immediately outside the villus capillaries, C_b, can be calculated, i.e.:

$$C_b = \frac{(C_V e^{PS/F} - C_{art})}{(e^{PS/F} - 1)} \tag{2}$$

C_b may also be calculated from the mean capillary concentration of glucose, C_m, since $C_b = C_m + J_S/PS$. There are several ways of estimating C_m, the simplest being to assume that glucose entry is constant throughout the transit of blood through the capillaries then C_m is the arithmetical mean of C_{art} and C_V.

The fall in concentration along the LIS, $C_a - C_b$, is determined by convection and diffusion. During glucose absorption, the bodies of the epithelial cells beneath the tight junctions becomes truncated cones so that the LIS widen progressively towards the basement membrane. Consequently the

Péclet number for convective and diffusive transport (essentially the ratio of the velocity of convection to that of diffusion) falls as the LIS enlarges. If the dimensions of the LIS are known, the Péclet number, Pe, can be calculated as:

$$Pe = \frac{J_V}{D} \int_{x=0}^{x=b} \frac{dx}{A(x)} \tag{3}$$

where J_V is net fluid flow through the LIS, D is the diffusion coefficient of glucose in the LIS, $A(x)$ is the cross-sectional area of the LIS at a distance x cm below the tight junction and $x = b$ is the value of x at the basement membrane. Once eq. (3) can be evaluated, the fall in glucose concentration down the length of the LIS, is calculated from:

$$(C_a - C_b) = \left(\frac{J_S}{J_V} - C_b \right)(1 - e^{-Pe}) \tag{4}$$

Evaluation of Equations (1–4)

Because it has been shown that glucose absorption rates are depressed by anesthesia [2], we have used eqs. (1–4) to analyze two sets of data from unanesthetized animals. The most extensive sets of data available to us were those from Thiry–Vella loops in unanesthetized rats [3–5] and data collected by several investigators from perfusion of jejunal segments in normal conscious human subjects [6–14]. These studies gave us values for J_S and J_V that were expressed respectively as µmol per hour ($\mu mol\,h^{-1}$) per cm^2 of smooth luminal surface ($cm^{-2}\,S_L$) and ml per hour ($ml\,h^{-1}$) per cm^2 of smooth luminal surface ($cm^{-2}\,S_L$). Pappenheimer [15] has shown that expressing values per unit area of the smooth luminal surface of the intestine is of great value in scaling intestinal absorption in mammalian species.

In addition to the fluxes, values for F, C_{art}, PS and the dimensions of the LIS during glucose absorption are also required before eqs. (1–4) can be evaluated but, apart from C_{art}, these are not available for unanesthetized rats and humans. We have therefore taken data from other species and used scaling functions [15] to obtain appropriate values. A full description of the sources of data and scaling calculations is given in Pappenheimer and Michel [1]. Here, it is important to draw attention to the increases in F and PS that occur during absorption. There are few data defining how F and PS increase with J_S [16–18] so we have assumed the increases are linear. Thus for villus blood flow, F, we have used the relations:

In rats: $F = 0.11 + 0.018 J_S$ (5a)

In humans: $F = 0.9 + 0.02 J_S$ (5b)

Similarly, we have expressed villus capillary PS as a linear function of F [19]. Thus

for rats: $\mathrm{PS} = 0.43 + 0.44F$ (6a)

and for humans: $\mathrm{PS} = 1.8 + 0.44F$ (6b)

In eqs. (5) and (6), both F and PS are expressed in units of $\mathrm{ml\,h^{-1}\,cm^{-2}\,S_L}$.

Results and Discussion

Interstitial Glucose Concentrations at the Epithelial Basement Membrane

Figures 2a and b show how C_b, the mean glucose concentration at the epithelial basement membrane immediately outside the villus capillary walls, increases with increasing rates of glucose absorption in rats and humans.

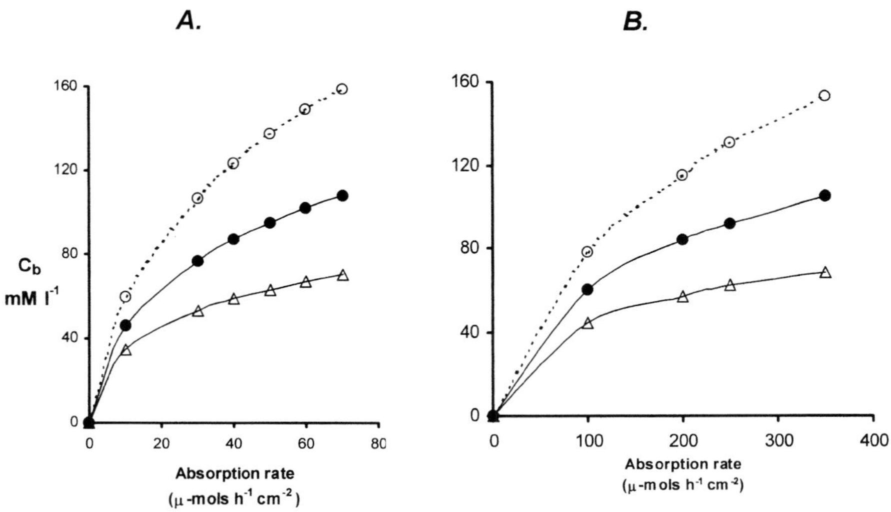

FIG. 2A,B. Predicted glucose concentrations at the epithelial basement membrane, C_b, just outside the villus capillaries are plotted against measured rates of glucose absorption. A shows values for rat and B shows values for healthy human subjects. The *solid circles* represent the changes when increments in villus blood flow (F) and permeability-surface area product (PS) with glucose absorption rate are "normal" [i.e., as in eqs. (5,6)]. The *open circles* show predictions when F and PS increments are 50% of normal and the *triangles* show values when increments are twice normal. From Pappenheimer and Michel [1], with permission

Because we are uncertain of the magnitude of the increase in F and PS with J_S, we show the changes in C_b with J_S calculated using eqs. (5) and (6) for values of F and PS and also those predicted when the increase in F is half as great and twice as great. The surprising result of these calculations is the very high level of glucose concentration even when the increases of both F and PS are twice as great as available evidence indicates. When eqs. (5a) and (5b) are used to estimate the increase of F with J_S, C_b rises above 100 mM at the higher rates of glucose uptake. If F increases only half as much as we have estimated, the rise in C_b is much greater. These very high concentrations of glucose in the absorbing villi have not been observed so far. They are, however, entirely consistent with the high tissue osmolality of the villi that has been reported in tissue taken during the absorption of salts and glucose [20,21]. They indicate that glucose is a major constituent of these hypertonic tissue fluids and that such high osmolalities can be achieved without the need for a counter-current multiplier system [20].

Although C_b is high it would be even higher if F failed to increase at the higher rates of glucose uptake. As seen in Fig. 2, the rate of increase of C_b with J_S diminishes with rising J_S and this is a result of the increase in F and consequent increase in PS with J_S. If after an initial increase, F and PS failed to rise further with increasing J_S, C_b would exceed 250 mM when $J_S = 70\,\mu\text{mol}\,\text{h}^{-1}\,\text{cm}^{-2}$ in rats and C_b would exceed 200 mM in humans when $J_S = 350\,\mu\text{mol}\,\text{h}^{-1}\,\text{cm}^{-2}$.

Not shown in Fig. 2 but emerging from the calculations are the gradients of glucose concentration across the walls of the villus capillaries. In rats, this rises from 18 to 68 mM as J_S rises from 10 to $70\,\mu\text{mol}\,\text{h}^{-1}\,\text{cm}^{-2}$. Similar concentration differences are seen across the villus capillary walls of human subjects i.e., from 32 to 70 mM, as J_S increases from 100 to $350\,\mu\text{mol}\,\text{h}^{-1}\,\text{cm}^{-2}$. These large concentration differences are necessary to account for the high fluxes that are observed in unanesthetized subjects.

Glucose Gradients Within the LIS

In rats, the differences in glucose concentration between the sub-junctional region, C_a, and the epithelial basement membrane, C_b, are small, rising to 3.7 mM when J_S is $70\,\mu\text{mol}\,\text{h}^{-1}\,\text{cm}^{-2}$. In humans, the higher glucose fluxes and the taller epithelial cells (with consequent longer LIS) give rise to larger differences between C_a and C_b. These rise from 5 to 20 mM as J_S increases from 100 to $350\,\mu\text{mol}\,\text{h}^{-1}\,\text{cm}^{-2}$. These calculations assume that all the glucose absorbed at the brush border flows along the entire length of the LIS, i.e., that all the glucose passing through the epithelial cells is extruded into the upper micrometer of the LIS just beneath the tight junctions. If glucose were extruded from the epithelial cells through a larger fraction of their lateral membranes bounding the LIS, C_a–C_b would be diminished further and become negligible in rats. The conclusion from these calculations is that compared with those occurring across the walls of the villus capillaries, the gradients of glucose in the LIS are small.

Glucose Concentrations in the Absorbing Epithelial Cells

Earlier we noted that glucose was carried into the epithelial cells with Na^+ on the SGLT-1 transporter and is carried out of the cell into the uppermost region of the LIS by facilitated diffusion on the Glut-2 transporter. From this, we know that the intracellular glucose concentration, C_{cell}, must exceed C_a in the uppermost region of the LIS to an extent that is determined by magnitude of glucose flux through the cells and also by the properties of the Glut-2 transporter. This suggests that C_{cell} might be estimated if the glucose flux through the cells and the properties of the Glut-2 transporter were known. Reliable estimates have been made for the fraction of the glucose flux that is carried on SGLT-1. In unanesthetized rats, Gromova and Gruzdkov [5] have shown this can be described by $48C_L/(C_L + 7)$ where C_L is the concentration of glucose in the gut lumen. In the steady state this should equal the flux of glucose through the lateral membranes into the LIS. The Glut-2 transporter obeys allosteric kinetics with a half saturation of 55 mM and a Hill coefficient of 1.6 [22]. The maximum transport capacity of Glut-2 in rats fed on a moderate carbohydrate diet has been determined by Cheeseman and Harley [23] to be 15 μmol min^{-1}(mg of epithelial cell protein)$^{-1}$. We have used these data to make tentative estimates of C_{cell} during glucose absorption in the rat [1].

In Fig. 3 we have expressed the results of our calculations in terms of the difference in concentration across the lateral cell membranes, $C_{cell} - C_a$, at different rates of glucose uptake. It is seen that while C_{cell} may exceed C_a by only a few mM at the lowest rates of glucose uptake, the concentration difference may rise from 14 to 54 mM as J_S increases from 30 to 70 μmol h^{-1} cm^{-2}. A striking prediction is that when the rate of increase of blood flow with J_S is halved, C_{cell} has to increase more rapidly than C_a to maintain the glucose flux. To maintain the highest rates of glucose transport, C_{cell} has to rise 170 mM above C_a. When one recalls that C_a is itself about 170 mM, C_{cell} must rise to 340 mM. Levels such as these might be expected to compromise the efficiency of the SGLT-1 upon which all glucose absorption depends. While these estimates of C_{cell} are based on very limited evidence, large increases in C_{cell} are likely to occur secondary to rises in C_a. At high absorption rates, the non-linear kinetics of Glut-2 become more obvious, and larger and larger values of $C_{cell} - C_a$ are required to maintain the efflux of glucose on Glut-2 as C_a rises. It would seem that this could be avoided only if V_{max} for Glut-2 were to increase with J_S.

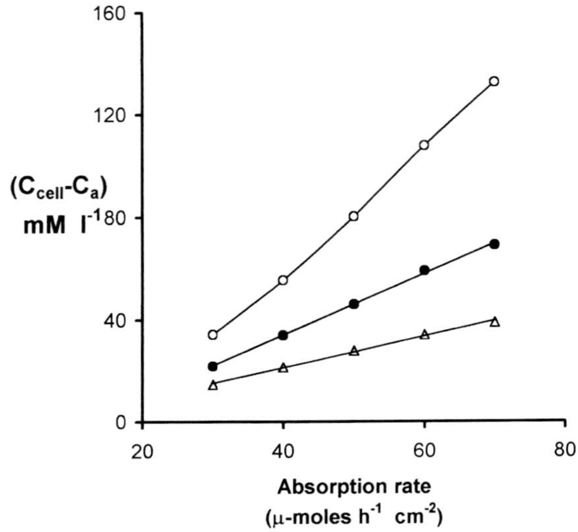

FIG. 3. Glucose concentration differences across the lateral membranes of rat jejunal epithelial cells ($C_{cell} - C_a$) predicted for different rates of glucose absorption. The *solid circles* show differences when *F* and PS increase by the "normal" extent with increasing glucose absorption rate. *Open circles* and *triangles* indicate changes when the increases in *F* and PS are 50% and 200% "normal," respectively. From Pappenheimer and Michel [1], with permission

Conclusions and Summary

When the best available estimates of the blood flow and the permeability-surface area product of the villus microcirculation are used in an analysis of glucose absorption in humans and rats, the concentrations of glucose predicted to be present at the basement membrane of the epithelium rise to values that exceed 100 mM. If an increase in villus blood flow did not occur in proportion to the glucose absorption rate, glucose concentration in the tissue would rise to even higher levels. Our tentative estimates of glucose concentration within the epithelial cells suggest that without increases in blood flow and PS of the villus microcirculation the rising levels of intracellular glucose concentration would limit maximal rates of absorption. It would seem that in addition to events occurring in the epithelial cells, the absorption of glucose involves a co-ordinated microvascular response comparable to that occurring in skeletal muscle during exercise.

References

1. Pappenheimer JR, Michel CC (2003) Role of the villus microcirculation in intestinal absorption of glucose: coupling of epithelial with endothelial transport. J Physiol 553:561–574
2. Ugolev AM (1987) Membrane transport and hydrolytic enzymes under physiological vs. acute experimental conditions. News Physiol Sci 2:186–190
3. Grudzkov AA (1993) Modern concepts of substance transfer across the pre-epithelial layer of the small intestine (in Russian). Sech J Physiol 79:19–32
4. Grudzkov AA, Gromova LV (1995) The coupling of disaccharide hydrolysis with absorption of released glucose in the small intestine in vivo (in Russian). Dokl Akad Nauk 342:830–832
5. Gromova LV, Grudzkov AA (1999) Hydrolysis-dependent absorption of disaccharides in the rat small intestine. Gen Physiol Biophys 18:209–224
6. Holdsworth CD, Dawson AM (1964) The absorption of monosaccharides in man. Clin Sci 27:371–379
7. Malawer SJ, Ewton M, Fordtran JS, et al (1965) Interrelationships between jejunal absorption of sodium, glucose and water in man. Am Soc Clin Invest 44:1072–1073
8. Gray GM, Ingelfinger FJ (1965) Intestinal absorption of sucrose in man: the site of hydrolysis and absorption. J Clin Invest 44:390–397
9. Gray GM, Ingelfinger FJ (1966) Intestinal absorption of sucrose in man: interrelation of hydrolysis and monosaccharide product absorption. J Clin Invest 45:388–397
10. Fordtran JS, Saltin B (1967) Gastric emptying and intestinal absorption during prolonged severe exercise. J Appl Physiol 23:331–335
11. Sladen GE, Dawson AM (1969) Interrelationships between the absorption of glucose, sodium and water by the normal human jejunum. Clin Sci 36:119–132
12. Fordtran JS (1975) Stimulation of active and passive sodium absorption by sugars in the human jejunum. J Clin Invest 55:728–737
13. Gisolfi CV, Summers RW, Schedl HP, et al (1992) Intestinal water absorption from select carbohydrate solutions in humans. J Appl Physiol 73:2142–2150
14. Fine KD, Santa Ana CA, Porter JL, et al (1993) Effect of D-glucose on intestinal permeability and its passive absorption in human small intestine in vivo. Gastroenterology 105:1117–1125
15. Pappenheimer JR (1998) Scaling of dimensions of small intestines in non-ruminant eutherian mammals and its significance for absorptive mechanisms. Comp Biochem Physiol A 121:45–58
16. Shoemaker WC, Yanof HM, Turk IM III, et al (1963) Glucose and fructose absorption in the unanesthetized dog. Gastroenterology 44:654–663
17. Sidery MB, MacDonald IA (1994) The effect meal size on the cardiovascular responses to food ingestion. Br J Nutr 71:835–848
18. Gallavan RH, Chou CC, Kvietys PR, et al (1980) Regional blood flow during digestion in the conscious dog. Am J Physiol 238:H220–H225
19. Perry MA, Granger DN (1981) Permeability of intestinal capillaries to small molecules. Am J Physiol 241:G24–G30
20. Lundgren O (1984) Microcirculation of the gastrointestinal tract and pancreas. In: Renkin EM, Michel CC (eds) Handbook of physiology, section 2, vol. IV. Part ii. American Physiological Society, Bethesda, pp 799–863

21. Hallbäck DA, Jodal M, Mannischeff M, et al (1991) Tissue osmolality in intestinal villi of four mammals in vivo and in vitro. Acta Physiol Scand 143:271–277
22. Kellett JL (2001) The facilitated component of intestinal glucose absorption. J Physiol 531:585–595
23. Cheeseman CI, Harley B (1991) Adaptation of glucose transport across rat enterocyte basolateral membranes in response to altered dietary carbohydrate intake. J Physiol 437:563–575

Oxygen Transport in the Microvessel Network

Kazuo Tanishita[1], Kazuto Masamoto[1], Tomoko Negishi[1], Naosada Takizawa[2], and Hirosuke Kobayashi[2]

Summary. Oxygen delivery in the brain tissue is carried out by a diffusion process principally determined by spatial differences of partial pressure of oxygen (pO_2). Previous studies identified inhomogeneous distribution of cerebral tissue pO_2. This inhomogeneous pO_2 distribution might be related to spatial variations in microvascular structure, because a large amount of oxygen is supplied from microvascular network. In this study, to evaluate the oxygen transport in the cerebral cortex, we focused on regional structure of microvascular network and pO_2 distribution in the rat somatosensory cortex. To this end, firstly, we characterized local tissue pO_2 distribution by using an oxygen microelectrode. Secondly, we quantified three-dimensional microvascular structure by combining a traditional method for casting blood capillaries with quantitative analysis by using confocal laser-scanning microscope. Finally, the regional variations in oxygen transport were estimated by using numerical simulation of oxygen transport based on these experimental data (i.e., pO_2 distribution and microvascular structure).

Key words: Oxygen transfer, Cerebral cortex, Blood flow, Microvessels, Computer simulation

Introduction

The brain is a highly oxidative organ and its consumption rate of oxygen accounts for 20% of that of the whole body. This large consumption rate must be met by continuous supply of oxygen, because lack of oxygen rapidly causes

[1] Faculty of Science and Technology, Keio University, 3-14-1 Hiyoshi, Kohoku-ku, Yokohama, Japan
[2] School of Allied Health Sciences, Kitasato University, Center of Information Science, Kitasato University, 1-15-1 Kitasato, Sagamihara, Kanagawa, Japan

irreversible damage to central nervous system. Acute hypoxic episodes cause a certain pattern of regional damage [1]. The cerebral cortex (e.g., layers III, V, and VI) is one of the most susceptible regions to hypoxia, and damage to sensorimotor function is particularly severe in humans that survive hypoxic/ischemic episodes. However, little is known about whether oxygen transport in intracortical regions relates to such selective vulnerability to hypoxia.

In the cerebral cortex, anatomical, metabolic, and functional variations occur in local regions [2,3], indicating that these spatial variations might be related to the selective damage to hypoxia in the cerebral cortex. Therefore, we need to assess the oxygen transport in the cerebral cortex, and to consider the spatial variations in structure and function of the cerebral cortex.

In the present study, to evaluate the oxygen transport in the cerebral cortex, we firstly measured pO_2 distribution in the rat somatosensory cortex by using an oxygen microelectrode. Secondly, we quantified three-dimensional microvascular structure in the rat somatosensory cortex using a confocal laser-scanning microscope (CLSM). Thirdly, regional variations in oxygen transport in the rat somatosensory cortex were estimated by using numerical simulation of oxygen transport based on such experimental values of tissue partial pressure of oxygen (pO_2) distribution and microvascular structure.

Materials and Methods

Measurement of Tissue pO_2 Distribution

Six male Wistar rats (9 weeks old) weighing 280–290 g were used for these experiments. The procedure in detail is described in [4]. Animal protocols were approved by the Bio Ethics Committee of the Faculty of the Science and Technology, Keio University.

The oxygen microelectrode was constructed according to the procedure developed by Baumgärtl and Lübbers [5], which is described in detail in [4]. The diameter of the electrode tip ranged from 2 to 10 μm.

Tissue pO_2 was measured perpendicularly from the dorsal surface at depth intervals of a 0.02 mm by positioning the microelectrode with a micromanipulator (ME-71, Narishige Scientific Instrument Lab, Tokyo, Japan). Location for the pO_2 measurement was selected based on a published brain map [6]. The position at which the microelectrode tip first contacted the cerebral surface was defined as 0 mm depth. Oxygen current was measured by using a micro-ammeter (R8340A, Advantest, Tokyo, Japan) with the voltage of 0.65 V applied. The recorded current was converted into pO_2 values by using the calibration curve. The average pO_2 is represented here as mean ± SD. Statistical significance was determined by a Student's *t*-test and set at $P < 0.05$.

Microvascular Density

The reconstructed images of the rat primary somatosensory cortex were separated into four major cortical areas (BF: barrel field; FL: forelimb region; Tr: trunk region; and HL: hindlimb region) and identified with a published brain map [6]. To calculate microvascular density, we extracted rectangular samples in each cortical area with perpendicular to the brain surface from the reconstructed image. We determined the microvascular density profile in each cortical area by calculating the number of black pixels (i.e., the number of vessels) parallel to the brain surface at each depth (each pixel was 1.43 μm) through the entire length of the extracted samples and then dividing the sum of the number of vessels at each depth by the width (in millimeters) of the sample. Consequently, successive depth profile of the microvascular density (in number of vessels per millimeter) was established at 1.43-μm (one pixel) intervals.

Numerical Simulation of Oxygen Transport

Based on the confocal microscopic images of the cerebral microvascular network, we calculated three-dimensional pO_2 distribution using flow and mass transfer modeling software (FLUENT; Fluent, Lebanon, NH, USA). Flow rate in the arteriole side was assigned in accordance with Poiseuille's law, the Hill equation was applied to consider the oxyhemoglobin dissociation curve in blood, and the facilitated diffusion coefficient was also incorporated. For the boundary conditions between tissue and vascular wall, equivalent pO_2 and oxygen flux were assigned, as described in Table 1. Figure 1 illustrates the model for superficial and middle layers, and the geometries of arteriole and capillary were determined by microscopic measurements using a confocal laser scanning microscope and micro computed tomography (CT).

TABLE 1. Assumptions for the calculation

- The oxygen gradient at the edge of module unit is zero
- The oxygen consumption rate is zero in the superficial layer up to 100 mm depth
- Matching flux at the vessel wall

	Superficial layer	Middle layer
Arteriole inlet Po_2	100 mmHg	85 mmHg
Capillary inlet Po_2	50 mmHg	50 mmHg
CBF (cm^3/100 g/min)	153	247

CBF, cerebral blood flow

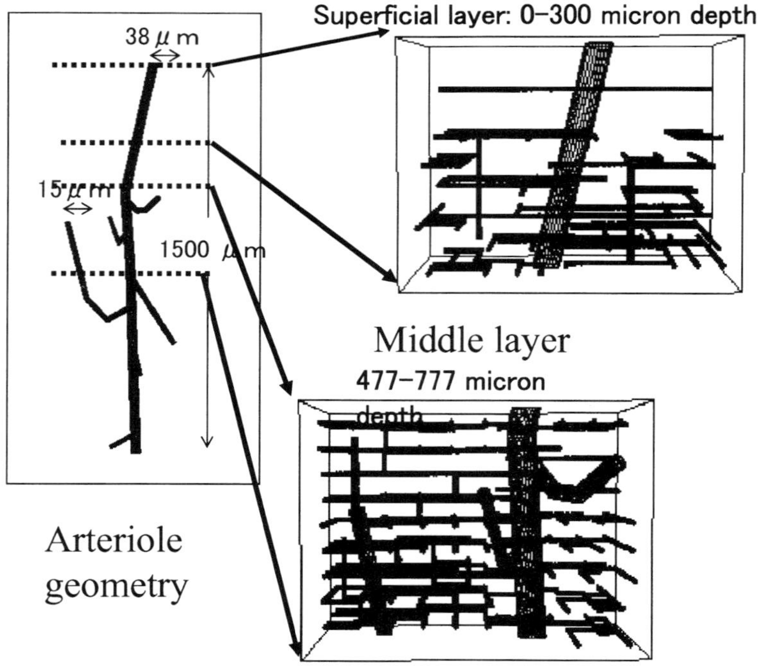

FIG. 1. Geometries of arteriole and capillary for superficial and middle layers

Results

Depth Profile of Tissue pO_2 Distribution

Average pO_2 profiles of all measurements revealed areal differences in local pO_2 among adjacent cortical areas (HL, FL, and Tr) in rat somatonsensory cortex, as shown in Fig. 2. The large error bars at each depth (0.02-mm intervals) indicate spatial variation among single trials in different measurements. In contrast, temporal change of pO_2 in each position was less than about 10% of the average pO_2 at a single position for all three areas. Comparison between HL and FL (Fig. 2A and B) shows that their pO_2 profiles were similar, wheaeas the pO_2 profile for the Tr (Fig. 2C) shows a relatively constant yet low average pO_2. Correspondingly, histological structure did not differ between the HL and FL, whereas the Tr had significantly thinner layers IV–VI compared with the same layers in the HL and FL (Fig. 2). Although no sign of tissue hypoxia specific to the respective layers was observed, comparison of all of the measurements reveal that the average pO_2 in the Tr (14 ± 10 torr) significantly differed from that in the HL (25 ± 13 torr) and FL (24 ± 13 torr).

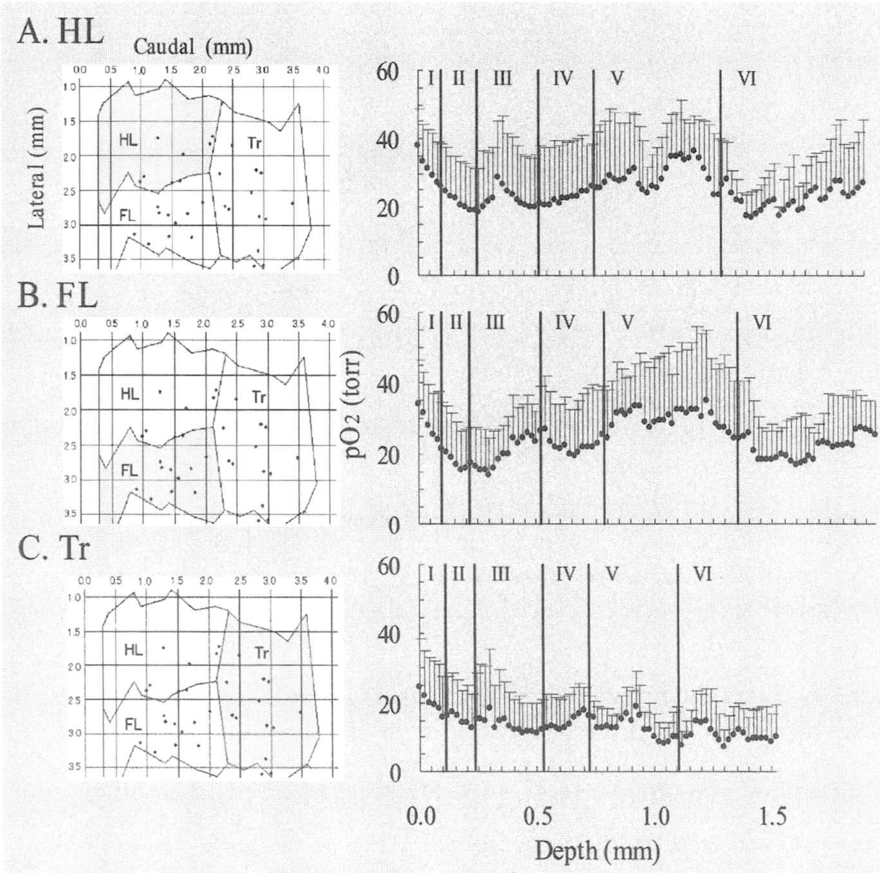

FIG. 2. Depth profiles of tissue pO_2 distribution for somatosensory areas hind limb (*HL*), fore limb (*FL*), and trunk (*Tr*)

Figure 3 shows a typical profile of average microvascular density at each depth in the BF from all 176 samples. This figure reveals specificity of successive depth variation in microvascular density in the intracortical region. Although each individual density profile had many sharp peaks, this average density profile showed no such peaks because the peaks in microvascular density (horizontal branches of microvessels) did not depend on specific depths. The microvascular density significantly increased in layers I–III from the brain surface toward layer IV, and slightly decreased in layers V–VI to white matter. The average density profile also showed two characteristic plateau regions, i.e., at depths of 0.3–0.8 mm and 1.0–1.5 mm. The former

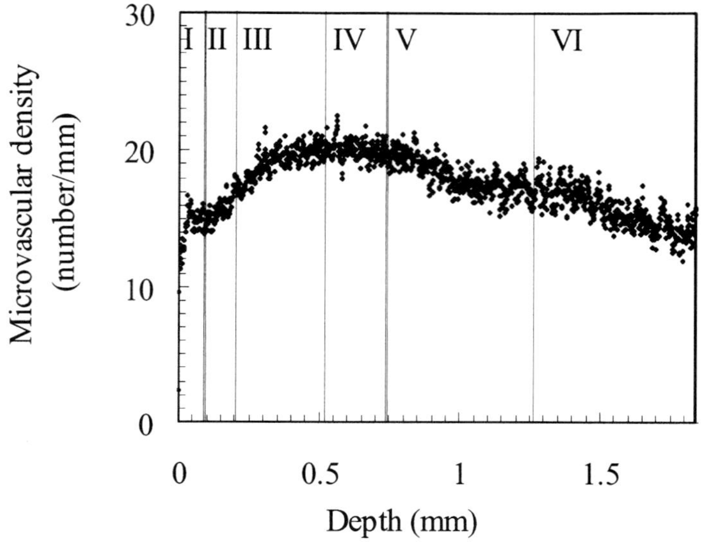

FIG. 3. Depth profile of capillary density

plateau in the middle layers (III–IV) was almost 50% higher in microvascular density than that in the superficial layers (I–II).

Contribution of Arteriole and Capillary to O_2 Supply

We calculated the oxygen supply from arteriole and capillary. Figure 4 shows the ratio of oxygen supply from capillary S_{cap} and from arteriole S_{art} as a function of cerebral blood flow (CBF) rate. If S_{cap}/S_{art} becomes larger, capillary contribution becomes larger. In both layers, when only arteriole CBF increased by 30%, S_{cap}/S_{art} slightly dropped due to the increased supply from arteriole. However, the S_{cap}/S_{art} increases with the increase of capillary CBF. Overall, the contribution of capillary was higher in the upper layer than in the middle layer. In contrast, the contribution of arteriole was higher in the middle layer than in the upper layer.

Discussion

Spatial Variations in pO_2 Distribution

Our results of spatial variations in pO_2 distribution reflect spatial variations in microvascular structure [7], microcirculation, and neuronal activity, because tissue pO_2 is determined by the oxygen content of blood, the rate of blood flow, and the rate of cellular oxygen consumption [8]. Although local

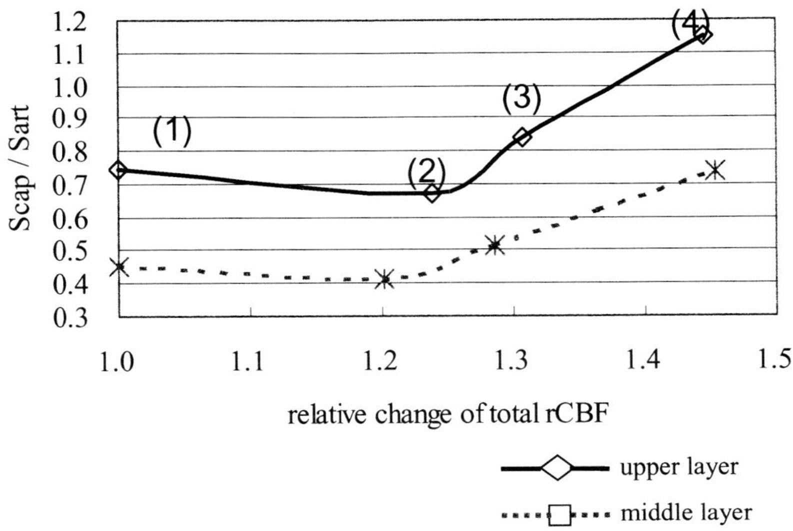

FIG. 4. Oxygen supply ratio as a function of cerebral blood flow (*CBF*) in the arteriole and capillary. Total rCBF parameter: (1) control flow; (2) arteriole CBF 30% up; (3) arteriole CBF 30% up + capillary CBF 33% up; (4) arteriole CBF 30% up + capillary CBF 100% up. *Scap*, oxygen supply from capillary; *Sart*, oxygen supply from arteriole

pO_2 temporally varied in some measurements, the temporal pO_2 changes were not significant because changes in tissue pO_2 during measurements of 10 s at the single position were small (<10% of average pO_2) in all trials. Such temporal changes in pO_2 were possibly caused by spontaneous vasomotion of localized small blood vessels [9]. In relation to the spatial variations in pO_2 distribution, we focused on cytoarchitectural differences between cortical areas because cytoarchitecture is closely related to cerebral function and metabolic activity. Consequently, a close correlation between cytoarchitecture and pO_2 distribution was revealed (Fig. 2). This result suggests that cerebral tissue pO_2 is strongly dependent on anatomical structure in the cerebral cortex.

We also identified pO_2 change due to the somatosensory stimulation [4] in the rat experiment. The pO_2 on the area of hindlimb and forelimb increased on stimulation, while pO_2 on the area of trunk decreased on stimulation. This may be explained by the regional variation of the balance between the oxygen demand and supply.

We should note the dynamic change of pO_2 for the nervous stimulation. We discovered a biphasic change of pO_2 in the cerebral tissue of hamster [10], and the tissue pO_2 in the activated region initially decreased during the 3 s after the onset of acoustic stimulation and then increased during the successive seconds. This may be explained by the delayed onset of the increase in cerebral blood flow, i.e., the initial decrease in tissue pO_2 is coupled to the induced neural activity and depends on the response time of the local increase in cerebral blood flow.

As we need to know how the blood flow contributes to the supply of oxygen into the cerebral tissue, we performed the computer simulation of oxygen profile. These considerations led us to investigate the role of arteriole and capillary networks for the oxygen transfer to tissue. In the middle layer, O_2 supply from arteriole was effective, because penetrating arteriole has some branches at this depth and distributes the blood flow into tissue. On the contrary, oxygen supply from capillary networks in the upper layer comes from the adjacent module or bottom layer.

Therefore, in neuronal activity the cerebral tissue may facilitate the effective O_2 transfer mechanism and necessary oxygen is transported to the localized region of high O_2 demand by arteriole, and the upper layer receives the oxygen through capillary networks from the nearby cortex.

References

1. Cervos-Navarro J, Diemer NH (1991) Selective vulnerability in brain hypoxia. Crit Rev Neurobiol 6(3):149–182
2. Wree A, Zilles K, Schleicher A (1990) Local cerebral glucose utilization in the neocortical areas of the rat brain. Anat Embryol (Berl) 181(6):603–614
3. Mountcastle VB (1997) The columnar organization of the neocortex. Brain 120(4): 701–722
4. Masamoto K, Takizawa N, Kobayashi H, et al (2003) Dual responses of tissue partial pressure of oxygen after functional stimulation in rat somatosensory cortex. Brain Res 979:104–113
5. Baumgart LH, Lubbers DW (1983) Microcoaxial needle sensor for polarographic measurement of local O_2 pressure in the cellular range of living tissue. Its construction and properties. In: Gnaiger E, Forstner H (eds) Polarographic oxygen sensors: aquatic and physiological applications. Springer, Berlin Heidelberg New York, pp 37–65
6. Paxinos G, Watson C (1998) The rat brain. Academic, Tokyo
7. Masamoto K, Kurachi T, Takizawa N, et al (2004) Successive depth variations in microvascular distribution of rat somatosensory cortex. Brain Res: 995:66–75
8. Popel AS (1989) Theory of oxygen transport to tissue. Crit Rev Biomed Eng 17(3): 257–321
9. Karkan DM, van Breemen C, Skarsgard PL, et al (1999) A link between vasomotion and spontaneous oscillations of oxygen in rat brain. Adv Exp Med Biol 471:111–116
10. Masamoto K, Omura T, Takizawa N, et al (2003) Biphasic changes in tissue partial pressure of oxygen closely related to localized neural activity in guinea pig auditory cortex, J Cereb Blood Flow Metab 23:1075–1084

Beating-Heart Microvascular Imaging by High-Speed Video Microscope and SPring-8

Fumihiko Kajiya

Summary. Flow velocity waveforms of coronary arterial inflow and venous outflow of myocardium are influenced by cardiac contraction and relaxation: unlike other organs arterial flow is exclusively diastolic; venous outflow is systolic. We first present some results of visualization of vascular image with flow in transmural microvessels by our needle-lens probe charge coupled device (CCD) microscope and a synchrotron radiation method. Then, an interpretation of the arteriolar and venular hemodynamics through a cardiac cycle was made by referring an intramyocardial vascular model with variable resistances, capacitances (ordinary capacitance and unstressed volume: UV), and an intramyocardial pump. After describing a hierarchical system of coronary microvessels, we emphasize the importance of capillaries as oxygen supply to myocytes, the function of capacitance (UV), and interaction with red blood cells through the glycocalyx.

Key words. Coronary microcirculation, Coronary slosh phenomenon, Functions of coronary capillary, Needle-lens probe CCD microscope, Synchrotron radiation method

Introduction

The phasic flow in the left coronary artery is diastolic-predominant. Scaramucci (1695, cited by Porter [1]) hypothesized that the deeper coronary vessels are squeezed by the contraction of the muscle fibers around them, which displaces the intramyocardial blood into coronary veins, and the vessels are refilled from the aorta during diastole. This hypothesis is certainly true:

Department of Cardiovascular Physiology, Okayama University Graduate School of Medicine and Dentistry, 2-5-1 Shikatacho, Okayama 700-8558, Japan the experimental measurements support that arterial blood inflow to the myocardium is almost exclusively diastolic, whereas venous outflow is predominantly systolic [2,3]. The flow characteristics suggest a substantial phasic volume change in the intramyocardial vessels by myocardial contraction and relaxation during a cardiac cycle, indicating importance of intramyocardial capacitance vessels. Thus, observation of the dynamics of intramyocardial microvessels during cardiac cycles is crucial in understanding the basic mechanism of intramyocardial blood perfusion. Mechanical stresses acting on intramyocardial blood vessels are different from epicardium to endocardium, so studies of coronary hemodynamics at different transmural depth provide important information about intramyocardial influence on blood distributions in the myocardial wall, giving the answer as to why the deeper portion of myocardium is vulnerable to ischemia, a principal question in coronary pathophysiology.

In the following, after mentioning the characteristics of coronary arterial and venous flow in small vessels which reflect the intramyocardial inflow and outflow, we describe (1) transmural microvascular dynamic imaging with blood flow in relation to cardiac contraction and relaxation, (2) the major vascular segment functioning as intramyocardial capacitance vessels, and (3) the functions of coronary capillaries. These studies were performed by means of our intravital microscope, a synchrotron radiation system (SPring-8, the world's most intense collimated beam of monochromatic X-ray with 8 GeV) and a confocal laser scanning microscope.

Intramyocardial Coronary Blood Flow During Cardiac Cycle

As mentioned above, coronary vessels in the myocardium are subjected to the phasic mechanical influences of cardiac contraction, resulting in a unique instantaneous blood flow pattern. Coronary arterial flow measured at just before its penetration into myocardium by our laser Doppler method exhibits a predominantly diastolic pattern, while venous flow a systolic pattern (Fig. 1) [3–5]. Thus, arterial inflow into myocardium during diastole should be stored in intramyocardial capacitance vessels, and an almost equal amount of blood should be squeezed out into epicardial veins in the next systole. This brings about a simple model of intramyocardial coronary circulation originally proposed by Spaan et al. and modified by us (Fig. 2) [6,7]. Unstressed volume (UV) is defined as a capacitance to accommodate blood during diastole without significant increase in pressure exceeding outflow venous pressure [7]. Thus, UV is important to explain the diastolic arterial inflow without venous outflow.

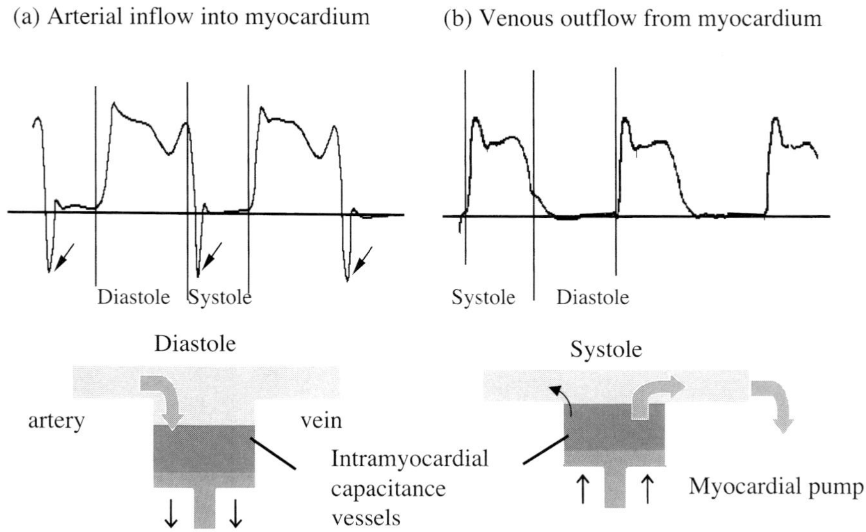

FIG. 1. Flow-velocity waveforms of arterial inflow (**a**) and venous outflow (**b**) of myocardium measured by laser Doppler blood flow velocimeter with an optical fiber. Coronary arterial flow is almost exclusively diastolic, while venous flow is systolic. Arterial reverse flow is recognized during early systole (*slanted arrows* in arterial flow tracing)

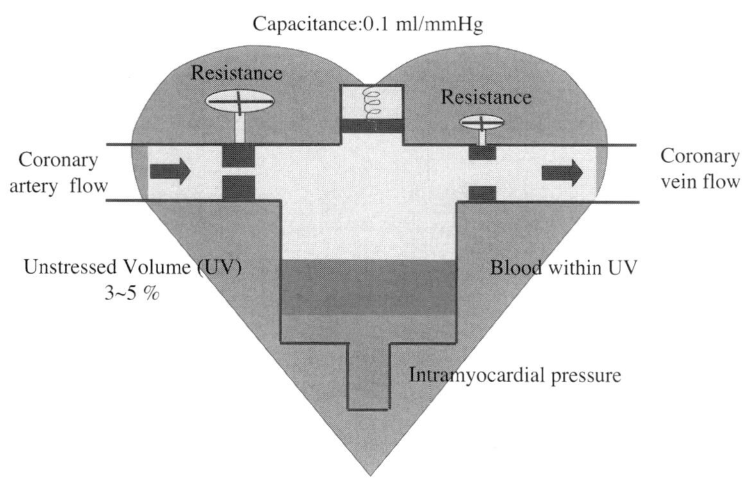

FIG. 2. Functional model of coronary circulation with variable resistances in arterial and venous sides, and intramyocardial capacitance (ordinary capacitance and unstressed volume). The variable resistances increase during systole and decrease during diastole, and unstressed volume (*UV*) accommodates arterial inflow during diastole without any increase in UV pressure above venous pressure. During systole an increase in intramyocardial pressure by cardiac contraction propels out the pooled blood in UV during diastole into veins. At the same time, a part of blood in UV is squeezed out into the artery as reverse flow (coronary slosh phenomenon). Unstressed volume accounts for 3%–5% of myocardial mass

In the coronary artery, a systolic retrograde blood flow is present when the instantaneous blood flow is studied in septal artery and/or a small epicardial artery just before penetration into the myocardium, indicating that a substantial amount of the blood stored in the myocardium during diastole returns to the proximal coronary arteries [3,8,9]. The forward and backward movement of the intramyocardial blood (coronary slosh phenomenon; see Fig. 1, slanted arrows) due to cardiac contraction and relaxation affects transmyocardial blood perfusion: it translocates intramyocardial blood from deeper to superficial myocardial layers during systole. Thus, the influence of systole on myocardial perfusion varied in different myocardial layers. The coronary slosh phenomenon is augmented when a coronary artery stenosis is present [10].

Transmural Intramyocardial Microvascular Hemodynamics During Cardiac Cycle

Subendocardial Microvascular Hemodynamics Visualized by a High-Speed, Needle-Lens Probe CCD Camera

To investigate the intramyocardial mechanical interaction between myocardium and coronary vessels and flow directly, we introduced a portable needle-probe video microscope with a charge-coupled device (CCD) camera to observe the subendocardial and intramural vessels (Fig. 3a) [11–13]. The phasic diameter of the intramural and subendocardial arterioles decreased 10%–20% by cardiac contraction (Fig. 3b,c, bottom; and see also Fig. 5). In contrast, the diameter of the subepicardial arterioles changes little during a cardiac cycle [11,12]. More recently, using a needle-probe microscope with a high-speed camera (200 frames/s), the movement of visible blood flow markers (blood velocities) exhibited exclusively diastolic flow in subendocardial arterioles (Fig. 3b, top) [14]. During systole, biphasic reverse flow components at early and mid-to-late systole appeared (Fig. 3c, top). These reverse flow components can contribute to the slosh phenomenon, causing the systolic retrograde coronary blood flow to perforating arterial branches and composing a part of systolic forward flow in subepimyocardial layers.

Observation of Hemodynamics of Perforating Branch by Synchrotron Radiation System (SPring-8)

Mori et al. observed the similar systolic narrowing of perforating arterial branches with length shortening by using an ordinary synchrotron radiation angiography in dogs [15]. The degree of systolic compression was high in deeper myocardium. They also detected to-and-fro movement of radio-

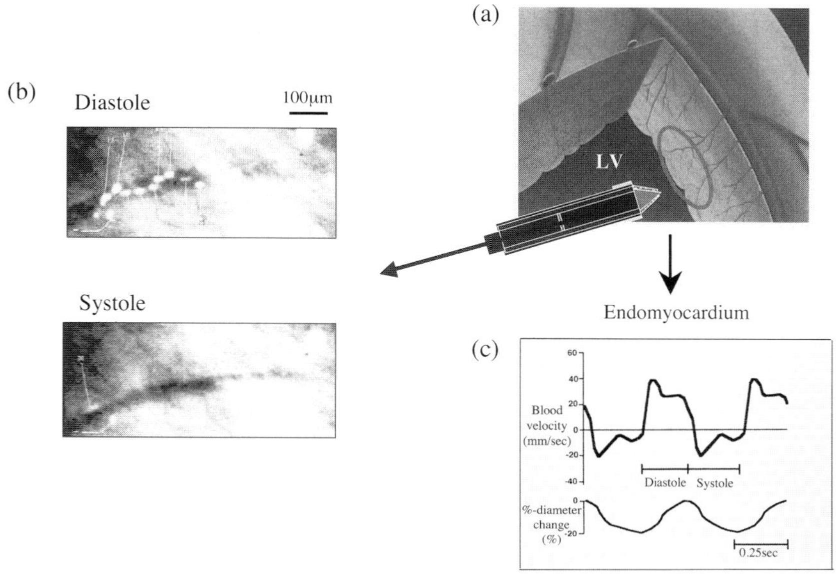

FIG. 3a–c. Phasic nature of diameter and flow in subendocardial arteriole visualized with our high-speed, needle-lens probe CCD microscope. **a** The portion of visualization, i.e., subendocardium. **b** Diastolic (*upper*) and systolic (*lower*) images of subendocardial arteriole with flow markers (*white particles* in the vessel). Note the systolic narrowing of the arteriole and the flow preponderance during diastole (markers seen almost exclusively during diastole). **c** *Upper trace*: diastolic predominant flows with systolic reverse flow. *Lower trace*: the diameter change during a cardiac cycle. Revised from Kajiya et al. [14], with permission

opaque material indicating the coronary slosh phenomenon there. Recently, we have succeeded in visualizing rat coronary microcirculation with iodine contrast agent by a 2-ms shutter with 33-KeV X-rays in SPring-8 [16]. Figure 4 shows a coronary microangiogram of a blood-perfused isolated rat heart during diastole. The perforating branches exhibited systolic compression in rats as seen in dogs.

Transmural Consideration of Coronary Microvascular Hemodynamics

It is generally accepted that intramyocardial pressure (IMP) during systole acts on intramyocardial microvessels (see Fig. 2) and decreases linearly from endocardium to epicardium, although the definition of IMP is difficult. This conjecture is compatible with observed intramural microvessels behaviors by inserting the needle-lens probe into myocardium [12]. Transmural arterial reverse flow (and venous forward flow) to epimyocardial layers during systole

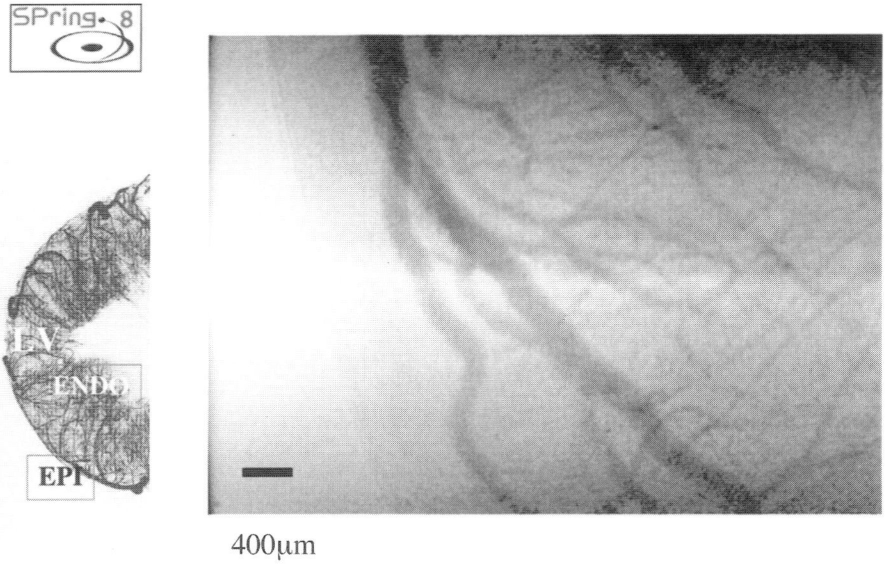

FIG. 4. Micro X-ray image of perforating arterial microvessels in a rat taken by SPring-8 synchrotron radiation. Perforating microvessels are clearly visualized by a high-speed shutter and high-resolution X-ray system. The degree of systolic compression (in video, not shown here) is similar to that of the intramyocardial arterioles shown in Fig. 3. *LV*, left ventricle; *EPI*, epimyocardium; *ENDO*, endomyocardium. Revised from Matsumoto et al. [16], with permission

is mainly from endomyocardial layers and partly from midmyocardial layers, so the intravascular pressure should be highest in the deeper, median in the middle, and lowest in the superficial layers (Fig. 5). Because of similar systolic narrowing of arterioles in deeper and middle layers (both horizontal arterioles and perforating arteries, Fig. 5), transvascular pressure should be almost constant from end to mid myocardial layers [12]. In other words, the relative balance between intramyocardial and intravascular pressures should not change from subendocardium to midcardial layers, i.e., both pressures should decrease in similar degree from deeper to middle layers to keep transvascular pressure constant through these layers. From mid to epicardial layer, however, the decrease in IMP may be greater than the decrease in intravascular pressure, causing little or no systolic narrowing of arterioles in epimyocardium.

In vivo observation of the intramyocardial venules demonstrated that the diameter in the midcardial venules was almost unchanged from diastole to systole (Fig. 5) [12]. The diameters of subendocardial venules decreased (10%–20%) from diastole to systole, whereas the diameter of the subepicar-

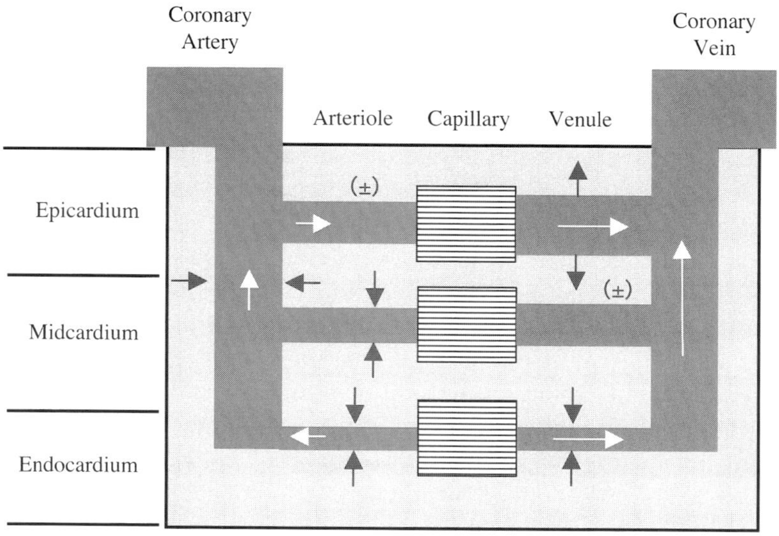

FIG. 5. Schematic illustration of diameter changes from diastole to systole and directions of blood flow of subepicardial, midcardial, and subendocardial microvessels during systole. *White arrows* indicate flow direction. *Black arrow-headed* vessels, 10%–20% decrease; (±), little change in diameter; *black arrow-counterheaded* vessels, 10%–20% increase in diameter

dial venules increased (10%–20%) [11,12]. The smaller intravascular pressure drop, i.e., almost uniform pressure from deep to superficial myocardium (due to low transmural venous resistances) relative to the larger IMP drop during systole explains the difference in the phasic venular diameter changes across the myocardium. Lower intravascular pressure relative to IMP in subendocardium during systole may cause systolic narrowing of the venules, whereas higher intravascular pressure relative to IMP in subepicardium during systole may result in systolic enlargement of the venules. The intramyocardial pressure and IMP in midcardial layer may be balanced, causing little venular diameter change.

Functional Role of Capillary as Capacitance and Unstressed Volume Through Myocardial Wall

To clarify the functional characteristics of intramyocardial microvessels, as well as possible transmural difference, we visualized the three-dimensional (3D) vascular architecture by intracoronary injection of $BaSO_4$, Indian ink, gelatin, and distilled water in diastolic and systolic-arrest rat hearts [17]. An X-ray microCT commercially available and SPring-8 microCT (the latter provides a finer image), and a confocal laser scanning microscope

(CLSM) were used for the 3D microvisualization. For microCT studies, transmural myocardial column (diameter 4 mm) was punched out from the left ventricle free wall and for CLSM studies, a 200-μm thickness block sample was used after slicing by a microtome.

Figure 6 shows arteriolar and venular images, and capillary images during diastole (left) and systole (right). The vascular volume change from diastole to systole in arterioles and venules was 48% (Figs. 6a and 7a) which roughly coincided with a 20% change in arteriolar diameter observed by the needle-lens probe microscope (see Fig. 3c, bottom). On the other hand, capillary volume decreased by 32% (Figs. 6b and 7b) during a cardiac cycle significantly smaller than the volume change in arterioles and venules. However, the volume fraction of capillaries per unit myocardial mass is ten times as large as that of arterioles and venules (Fig. 7), indicating that capillaries function as major capacitance and unstressed volume (see Fig. 2).

Transmurally, the reduction in volume fraction of capillary from systole to diastole was 37% in endocardium, 34% in midcardium and 19% in epicardium from the study by CLSM. Thus, the functions of capillary as capacitance may be more remarkable in the deeper layers.

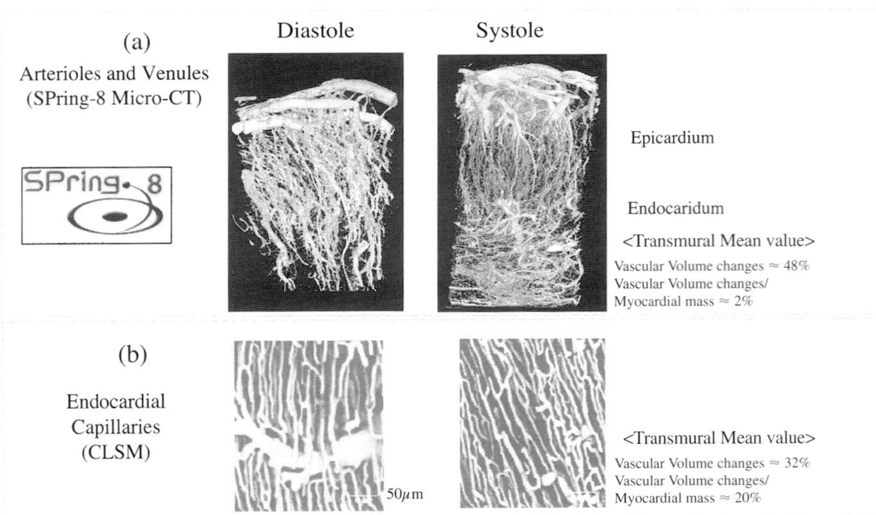

FIG. 6. Transmural arteriolar and venular images (**a**), and endocardial capillary image (**b**). *Left panels* of **a** and **b** indicate diastolic images, *right panels* systolic images. Note that arterioles, venules, and capillaries are compressed during systole, significantly decreasing their diameters. Revised from Toyota et al. [17], with permission

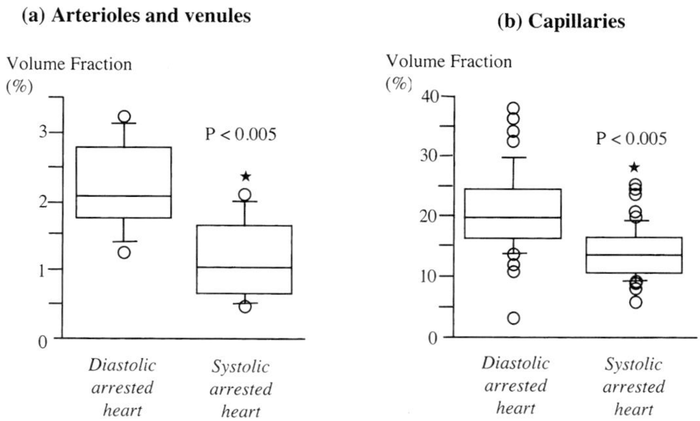

FIG. 7. Transmural fractional volume changes from diastole to systole in arterioles and venules (**a**) and capillaries (**b**). The percent volume change of arterioles and venules was 48%, while that of capillaries was 32%. However, the volume fraction per myocardial mass of capillaries was about ten times larger than arterioles and venules, indicating that capillaries mainly function as capacitance. Revised from Toyota et al. [17], with permission

Capillary Flow Dynamics and its Functional Role

Recently, we visualized the epicardial capillary network of beating rat hearts in vivo, using high-resolution needle-lens probe microscopy (Fig. 8). A capillary originating from a point in an arteriolar zone carries blood running parallel to the muscle fibers and ends in a point in the venular zone. The capillary length from the smallest arteriole and venule is several hundred micrometers [18]. This path length is 8–10 times larger than a single capillary length, indicating 8–10 interconnections along the path length. The flow through Y, T, H, and hairpin-type interconnection seen in the CLSM image (Fig. 6) was recognized by in vivo visualization.

The transit time of blood in the capillary path length was about 1.5 s, which was estimated from the resumption of venular flow after release of transient occlusion of coronary artery [19]. The existence of transit time indicates that the capillary works as unstressed volume which stores the arterial inflow blood for ~1.5 s without emergence of venous flow. Capillary flow was predominant either during systole or during diastole. The coexistence of temporal flow preponderance (systolic vs diastolic) implies that the watershed between diastolic arterial and systolic venous flows is located within capillaries. The diastolic preponderance may be arteriolar capillary, while the systolic may be a venular one.

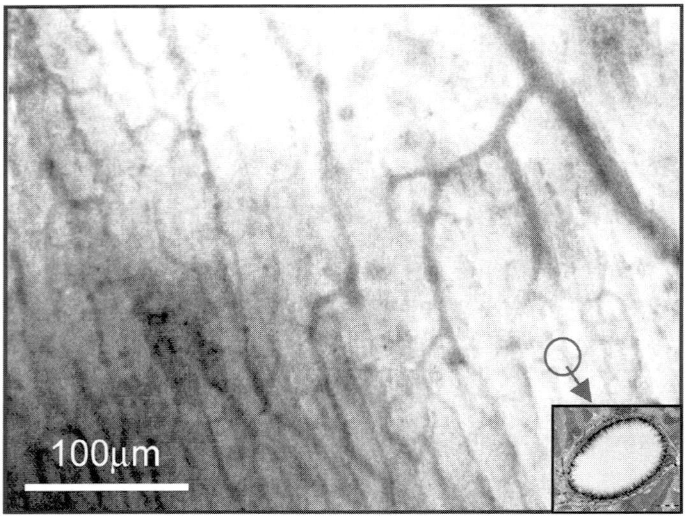

FIG. 8. Visualization of coronary capillary network in rats visualized with our high-magnification needle-lens CCD microscope. Note the dense capillary network with H, Y, T, and hairpin-type interconnection. *Inset*: glycocalyx layer in the capillary (courtesy of Drs. H. Vink and J. Spaan, University of Amsterdam)

Another interesting feature of capillary is the bush-like structure of the glycocalyx (Fig. 8, inset) as shown by van den Berg et al. [20]. Because red cells are continuously flowing through the capillary network with different velocities, the structural properties may be attributable to attenuated direct physical and chemical interaction between endothelial cells and red blood cells, facilitating smooth blood flow in capillary [21]. On the other hand, the bush-like glycocalyx on the endothelial cells may amplify a small shearing force on the cells other than capillaries. In fact, nitric oxide production in in vivo arteries by an application of shear stress became negligibly small following degradation of the glycocalyx layer [22].

Conclusion

Coronary microvessels play a crucial role for mechanoenergetic interaction between blood flow and myocardial function, which is not uniform transmurally. Thus, highly organized vascular regulations are required for matching local blood flow with myocardial energy requirements. Recently, new technologies to visualize in vivo coronary microcirculation with new knowledge of the signaling for vascular regulation have revolutionized our abilities to understand the integrative regulation of coronary microcirculation. In this paper, beating-heart microvascular imaging by needle-lens probe microscope and SPring-8 with a confocal laser scanning microscope was introduced, mainly focusing on their integrational roles in maintaining coronary microvascular function.

Acknowledgments. I thank Drs. M. Goto, O. Hiramatsu, T. Kiyooka, T. Matsumoto, Y. Ogasawara, H. Tachibana, E. Toyota, K. Tsujioka, and T. Yada for their collaboration in this study. Thanks are also given to Ms. H. Izushi, E. Nawachi, and K. Yoshioka for their help in the preparation of the manuscript. This study was partly supported by Grants-in-Aid for Scientific Research (13854030, 15650095) from the Ministry of Education, Science, Technology, Sports, and Culture, and the Research Grant for Cardiovascular Diseases (14-1) from the Ministry of Health, Labor and Welfare, and a Cardiac Physiome Grant from Okayama Prefecture New Technology Promotion Foundation.

References

1. Porter WT (1898) The influence of the heart-beat on the flow of blood through the walls of the heart. Am J Physiol 1:145–163
2. Chilian WM, Marcus ML (1982) Phasic coronary flow velocity in intramural and epicardial coronary arteries. Circ Res 50:775–781
3. Kajiya F, Tomonaga G, Tsujioka K, et al (1985) Evaluation of local blood flow velocity in proximal and distal coronary arteries by laser Doppler Method. J Biomech Eng 107:10–15
4. Kajiya F, Hoki N, Tomonaga G, et al (1981) A Laser-Doppler-Velocimeter using an optical fiber and its application to local velocity measurement in the coronary artery. Experientia 37(11):1171–1173
5. Nishihara H, Koyama J, Hoki N, et al (1982) Optical fiber laser Doppler velocimeter for high-resolution measurement of pulsatile blood flows. Appl Optics 21(10):1785–1790
6. Spaan JAE, Breuls NP, Laird JD (1981) Diastolic-systolic coronary flow differences are caused by intramyocardial pump action in the anesthetized dog. Circ Res 49(3): 584–593
7. Kajiya F, Tsujioka K, Goto M, et al (1986) Functional characteristics of intramyocardial capacitance vessels during diastole in the dog. Circ Res 58(4):476–485
8. Carew TE, Covell JW (1976) Effect of intramyocardial pressure on the phasic flow in the intraventricular septal artery. Cardiovasc Res 10:56–64
9. Eckstein RW, Moir TW, Driscol TE (1963) Phasic and mean blood flow in the canine septal artery and an estimate of systolic resistance in deep myocardial vessels. Circ Res 12:203–211
10. Kimura A, Hiramatsu O, Yamamoto T, et al (1992) Effect of coronary stenosis on phasic pattern of septal artery in the dog. Am J Physiol 262:H1690–H1698
11. Yada T, Hiramatsu O, Kimura A, et al (1993) In vivo observation of subendocardial microvessels of the beating porcine heart using a needle-probe videomicroscope with a CCD camera. Circ Res 72:939–946

12. Hiramatsu O, Goto M, Yada T, et al (1998) In vivo observations of the intramural arterioles and venules in beating canine hearts. J Physiol 509:619–628
13. Kajiya F, Yada T, Kimura A, et al (1993) Endocardial coronary microcirculation of the beating heart. Adv Exp Med Biol 346:173–180
14. Kajiya F, Yada T, Matsumoto T, et al (2000) Intramyocardial influences on blood flow distributions in the myocardial wall. Ann Biomed Eng 28:897–902
15. Mori H, Tanaka E, Hyodo K, et al (1999) Synchrotron microangiography reveals configurational changes and to-and-for flow in intramyocardial vessels. Am J Physiol 276:H429–H437
16. Matsumoto T, Tachibana H, Asano T, et al (2004) Pattern differences between distributions of microregional myocardial flows in crystalloid- and blood-perfused rat hearts. Am J Physiol Heart Circ Physiol 286:H1331–H1338
17. Toyota E, Fujimoto K, Ogasawara Y, et al (2002) Dynamic changes in three-dimensional architecture and vascular volume of transmural coronary microvasculature between diastolic- and systolic-arrested rat hearts. Circulation 105(5):621–626
18. Kassab GS, Fung YC (1994) Topology and dimensions of pig coronary capillary network. Am J Physiol 267:H319–H325
19. Kiyooka T, Hiramatsu O, Shigeto F, et al (2003) Functional role of capillaries in reactive hyperemia by direct observation with a pencil-lens intravital videomicroscope. Microcirc Annu (Tokyo) 19:63–64
20. van den Berg BM, Vink H, Spaan JA (2003) The endothelial glycocalyx protects against myocardial edema. Circ Res 92(6):592–594
21. Weinbaum S, Zhang X, Han Y, et al (2003) Mechanotransduction and flow across the endothelial glycocalyx. Proc Natl Acad Sci USA 100(13):7988–7995
22. Mochizuki S, Vink H, Hiramatsu O, et al (2003) Role of hyaluronic acid glycosaminoglycans in shear-induced endothelium-derived nitric oxide release. Am J Physiol 285(2):H722H726

The Laser Hot Balloon Catheter: a New Approach to Treating Atheromatous Lesions

Tsunenori Arai

This chapter describes a new type of thermal balloon angioplasty. Our laser hot balloon (LHB) catheter is characterized by short-duration heating (<10 s) with uniform temperature distribution throughout the balloon. We have also successfully developed independent control of balloon temperature and the heating duration. We used the new technique to study the optimum combination of balloon temperature and heating duration to suppress restenosis after percutaneous transluminal coronary angioplasty (PTCA).

PTCA has been the gold standard treatment for ischemic heart diseases, in particular for acute myocardial infarction and angina. Restenosis in the chronic phase after PTCA has been the most serious issue to date, with approximately 30%–40% of patients suffering another heart attack due to restenosis in the treated vessel. Hot balloon angioplasty involves softening the vessel wall to obtain sufficient dilatation by temperature elevation and the addition of localized thermal coagulation to suppress granulation tissue growth in the vessel. Using a radio-frequency (RF) hot balloon, sufficient expansion of the lumen at low dilatation pressure has been demonstrated; however, the problem of restenosis after PTCA has not been solved. The RF balloon requires 1 min to reach 60°C, allowing thermal injury to spread around the vessel and over the vessel wall.

We developed the new LHB catheter with a balloon 3 mm in diameter and 20 mm in length. One lumen of the LHB catheter is used for guidewire insertion and the other two for contrast medium irrigation with a thin silica fiber installed in one of the irrigation lumens, and a thin thermocouple in the other. The continuous-wave-diode laser beam is focused into the silica fiber, and a metal mesh tube is located at the tip of the fiber, which is connected to the

Faculty of Science and Technology, Department of Applied Physics and Physico-Informatics, Keio University, 3-14-1 Hiyoshi, Kohoku-ku, Yokohama, Kanagawa 223-8522, Japan irrigation lumen and placed into the balloon. The laser beam emerges from the tip of the fiber and is reflected many times in the metal mesh tube, converting the light to heat energy. This heat is transferred to the contrast medium, which is irrigated into the lumen. The heated contrast medium flows through the balloon to obtain a uniform temperature distribution and the contrast medium temperature in the balloon is measured with the thermocouple.

Design and Modification of Nanoparticles for Blood Substitutes

Shinji Takeoka

Key words. Liposome, Nanoparticle, Hemoglobin, Red-blood-cell substitute, Platelet substitute

Introduction

We have been developing the technology of stabilized and functionalized nanoparticles such as liposomes for 20 years. When phospholipids and cholesterol are dispersed into an aqueous solution, they spontaneously assemble to form vesicles with a bimolecular (bilayer) membrane. There are many parameters, such as size, size distribution, lamellarity (the number of bilayer membranes), membrane fluidity, surface charge, surface modification, membrane permeability, that characterize liposomes. They can be adjusted as need dictates to allow for changing encapsulation of functional molecules, release triggered by external stimuli, conjugation of functional sites on the surface, rolling or adhesion properties of liposomes, and control of blood circulation time. On the other hand, we have to consider their physical and chemical stability during storage or blood circulation. Surface modification with polyoxyethylene (POE) chains is one of the most effective ways to impart such stabilization.

In this chapter, I introduce two examples of nanoparticle application; one is a liposome encapsulating concentrated hemoglobin (Hb-vesicle) for a red-blood-cell substitute, and the other is a liposome bearing recognition proteins or peptides on the surface and used as a platelet substitute. The microcirculation, pharmacokinetics, and histopathological change were studied in relation to the characteristics of the particles as well as their oxygen-binding and releasing properties. In the case of the platelet substitutes, nanoparticles

---

Department of Applied Chemistry, School of Science and Engineering, Waseda University, 3-4-1 Ohkubo, Shinjuku-ku, Tokyo 169-8555, Japan bearing receptor proteins of the platelet surface recognize the collagen surface under shear rates. The rolling and adhesion properties will be discussed depending on the stiffness or membrane fluidity of the particles.

Present Status of the Development of Red-Blood-Cell Substitutes [1,2]

Hb-vesicles that encapsulate concentrated hemoglobin with a phospholipids bilayer membrane have a similar structure to red blood cells, and are expected to be used soon in clinical tests because the degree of safety and efficacy are considered to be high. Although effective use of the hemoglobin from donated and expired blood should be promoted at the present stage, recombinant human hemoglobin will be used in the future. During hemoglobin purification from red blood cells, stroma including the glycoproteins which determine a blood type, proteins other than hemoglobin, and the viruses are removed by heating or filter processing. By encapsulating hemoglobin with a stabilized phospholipids membrane with POE-lipid, liquid-state preservation for 2 years is guaranteed at room temperature under nitrogen atmosphere [3], and with dry powder, further prolonged preservation is possible. These points are advantages for an artificial oxygen carrier.

The design of the red-blood-cell substitutes (POE-modified hemoglobin vesicles) are summarized in Fig. 1. Hb-vesicles are dispersed into a saline solution and enclosed with the bottle in a state of deoxidization. The hemoglobin concentration is 10 g/dl and is close to that of human blood. Moreover, because hemoglobin molecules are encapsulated, the colloid osmotic pressure of the solution is zero. Therefore, when regulation of colloid osmotic pressure

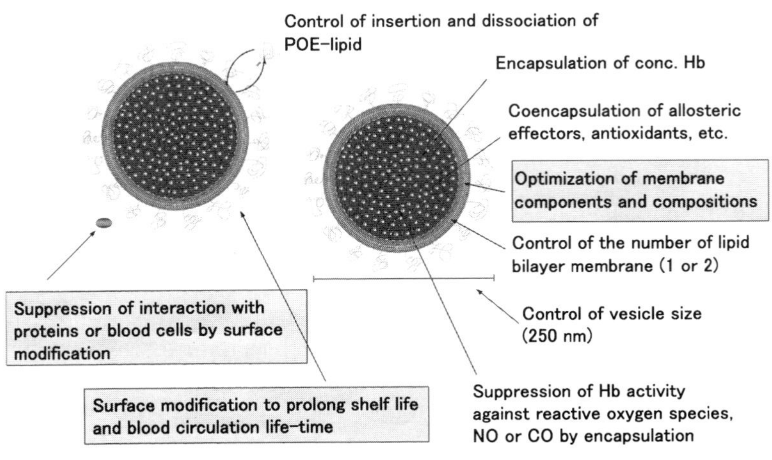

Fig. 1. Design of POE-modified hemoglobin vesicles (Hb-vesicles) as red-blood-cell substitutes

is needed, a solution of colloids such as albumin and polysaccharide will be used with the Hb-vesicle dispersion. The particle diameter is strictly adjusted to 250 nm. The degree of oxygen affinity is adjusted to a suitable value by coencapsulating an allosteric effector such as pyridoxal 5′-phosphate. The optimization of the composition of the lipid components resulted in high encapsulation efficiency of hemoglobin in the Hb-vesicle, a stability of 2 years in a liquid state, the prevention of hemolysis, an appropriate lifetime in blood circulation, and avoidance of platelet and complement activation. Furthermore, large-scale manufacturing has been improved by the introduction of freeze-thawing and freeze-drying operations which can control a molecular assembling state before encapsulating hemoglobin molecules.

Present Results of Safety and Efficacy Tests

Although in vivo testing was carried out using rats or hamsters, we confirmed the fundamental safety and oxygen transporting ability. Safety tests using primates is in progress. When 90% of the volume of rat blood was exchanged by the albumin dispersion of the Hb-vesicles, the oxygen partial pressure of the renal cortex was maintained as was blood pressure [4]. On the other hand, when the blood was exchanged by an albumin solution in the same concentration, the fall of blood pressure and oxygen partial pressure of the renal cortex became noticeable at 70% exchange, and all rats died just after 90% exchange.

In the hamster 80% exchange transfusion examination with the albumin dispersion of the Hb-vesicles, the noninvasively measured oxygen partial pressure of the subcutaneous tissue microcirculatory system was maintained at 5 times or more than that of the control albumin group although it fell to 60%–70% before exchange [5]. The contraction of a resistance blood vessel and the rise of blood pressure was not confirmed at all, but it was confirmed with modified hemoglobin products. Because the Hb-vesicle has a size that does not penetrate a blood vessel, there is no influence on the activity of nitrogen oxide as an endothelium-derived relaxation factor [6]. Furthermore, the Hb-vesicles cannot penetrate the sinusoidal vessels of liver (several holes 10–200 nm in size are open in the blood vessels) like old red blood cells, but are metabolized by Kupfer cells of liver and macrophages in reticuloendothelial systems. On the other hand, acellular hemoglobin molecules in the liver influenced liver microcirculation by eliminating carbon monoxide as a gaseous vasodilator, caused overgeneration of bilirubin, and suppressed bile secretion [7,8]. The half-life of Hb-vesicles in human blood circulation was estimated to be about three days on the basis of the results in rats, rabbits, and monkey. Moreover, from details of the blood biochemistry examination and pathology examination in the single and repetitive administration, we confirmed the transitional rise of lipase in connection with lipid decomposition, the transitional rise of a cholesterol value, iron content, and the transitional hypertrophy of a reticuloendothelial system [9]. It was concluded that Hb-vesicles are expected to function adequately in cases of extensive bleeding.

Development of Platelet Substitutes

The history of platelet substitutes is short compared with that of red blood cells, with few examples of research. A platelet has complicated functions, such as adhesion specific to the bleeding site, expansion, aggregation, secretion, and the activation of a blood coagulation system. Needless to say, we cannot make platelet substitutes that have these all. However, a bleeding tendency is strongly apparent in such bleeding diseases as Bernard-Soulier syndrome and thrombasthenia, in which adhesion and aggregation ability are lacking. In these conditions, a hemostatic effect can be expected by the infusion of particles having functions such as adhesion and the aggregation of platelets due to the assistance of the function of the remaining platelets. Although clinical tests were carried out with human red blood cells [10] or albumin microcapsules [11] conjugating fibrinogen, and with the dried powder of human platelets [12], all clinical tests were suspended due to problems of safety and efficacy. Moreover, since the blood components of human origin were used, the risk of infection cannot be avoided completely. The platelet substitutes created by our group use liposomes and recombinant human albumin as biocompatible particles. They also use recombinant proteins of the part of platelet membrane or synthetic oligopeptides by conjugating to those particles for the purpose of accumulation to the bleeding site involving native platelets, expecting to achieve hemostasis.

The mechanisms of platelet adhesion differ between the blood flow of high shear rate and that of low shear rate. As shown in Fig. 2, the hemostasis of the platelets to bleeding in a high shear rate begins from a platelet recognition of the von Willebrand factor (vWf) bound to the collagen in the subcutaneous tissue of a blood vessel exposed to the bleeding site, followed by platelet adhesion and rolling. This recognition ability comes from the GPIb/V/IX complex containing GPIbα on the surface of a platelet [13]. Next, the platelets will progress and a granule will be secreted if GPIaIIa ($\alpha_2\beta_1$ integrin) or GPVI on the surface of the platelet directly interacts with the collagen and is activated. Fibrinogen recognizes the activated GPIIb/IIIa ($\alpha_{IIb}\beta_3$ integrin), constructs a crosslink between blood platelets, forms an aggregate, and serves as primary hemostasis. Hemostasis is completed by formation (secondary hemostasis) of the fibrin clot by induction of the coagulation system.

The target platelet substitutes bear the water-soluble part of receptor proteins such as GPIbα which recognizes collagen through vWf under the blood

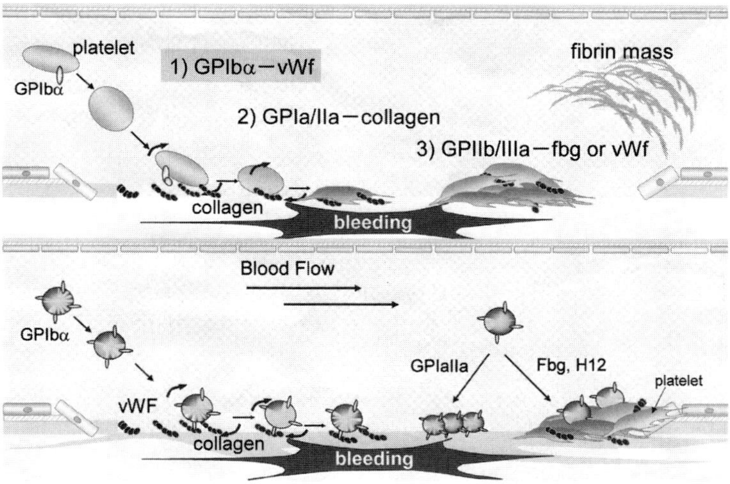

FIG. 2. Design of platelet substitutes studied from functions of natural platelets

flow of high shear rates and GPIaIIa which recognizes collagen directly at low shear rates. Instead of GPIIb/IIIa on the surface of an activated platelet, they bear the fibrinogen or its oligopeptide to assist the platelet aggregation as ligands of the GPIIb/IIIa.

If an anti-GPIIb/IIIa antibody is added to a platelet dispersion to inhibit GPIIb/IIIa activity, the platelets roll on the vWf-immobilized plate along the flow direction by the interaction of rGPIbα on the platelet and the vWf. Interestingly, phospholipid liposomes conjugating rGPIbα roll on the vWf-immobilized plate as well [14]. The number of the rolling liposomes increased with the shear rate, indicating the characteristic of rGPIbα. Moreover, the rolling speed was correlated with the membrane fluidity of the liposomes. That is, the rolling speed of "soft" liposomes was low, whereas that of the "hard" liposomes was high [15]. However, the rGPIbα-liposomes did not continue rolling but departed from the plate after rolling some length. This was remarkable as the "soft" liposomes. When the amount of rGPIbα on the surface of the liposome after the experiment was measured, it was suggested that rGPIbα-lipid should dissociate from the bilayer membrane during the rolling on the vWf-plate. Now rGPIbα-lipid which cannot dissociate serves as a point of a molecular design.

On the other hand, the rGPIaIIa-liposomes directly recognize collagen under the blood flow of low shear rates and adhered to (stopped at) the collagen-immobilized plate [16]. In this case, the number of the adhering liposomes decreased as the shear rate rose. However, liposomes conjugating both rGPIbα and GPIaIIa adhered on the collagen plate under the blood flow from low to high shear rates [17]. If the liposomes having platelet activation factors or coagulation factors in the internal aqueous phase accumulate at the bleeding site, they will be able to contribute effectively to hemostasis by releasing their contents. We also focused on using polymerized albumin particles as effective platelet substitutes and obtained some unique in vitro and in vivo results [18,19].

Conclusions

For red-blood-cell substitutes the present target is the supportive treatment of transfusion therapy in emergency, and nonclinical and clinical studies will be scheduled within 2 years.

On the other hand, the research of platelet substitutes has just started. There is a conflict between the carrier design for the extension of circulation lifetime and the carrier design to show the hemostatic activity by recognizing the bleeding site; therefore, we need to resolve this conflict and to design platelet substitutes for prophylactic or chronic treatments. And we also need a method to confirm that the candidate does not create a thrombus in blood circulation.

Although profitability is important in the development of blood substitutes, one that is based on a long-term view with consideration for human beings is expected first. At present, as many discoveries about the dynamic function of platelets have accumulated in a short period of time with the progress of biotechnology and opto-electronics, and the manufacturing technology of recombinant proteins or carriers is progressing, a usable product is sure to be invented in the near future.

Acknowledgments. The author thanks Drs. E. Tsuchida, H. Sakai, K. Sou, Y. Teramura, and Y. Okamura at Waseda University, and Drs. K. Kobayashi, Y. Ikeda, M. Suematsu, H. Horinouchi, M. Handa, and M. Murata at Keio University for useful discussions and suggestions. This work was supported in part by Health and Labor Sciences Research Grants (Research on Pharmaceutical and Medical Safety); the Ministry of Health, Labor, and Welfare, Japan; and Grants-in-Aid from the JSPS, Japan, and 21COE "Practical Nano-Chemistry" from MEXT, Japan.

References

1. Tsuchida E (ed) (1995) Artificial red cells: materials, performances and clinical study as blood substitutes. Wiley, Chichester
2. Tsuchida E (ed) (1998) Blood substitutes: present and future perspectives. Elsevier, Amsterdam

3. Sakai H, Tomiyama K, Sou K, et al (2000) Poly(ethylene glycol)-conjugation and deoxygenation enable long-term preservation of hemoglobin-vesicles as oxygen carriers in a liquid state. Bioconjug Chem 11:425–432
4. Sakai H, Takeoka S, Park SI, et al (1997) Surface modification of hemoglobin vesicles with poly(ethylene glycol) and effects on aggregation, viscosity, and blood flow during 90% exchange transfusion in anesthetized rats. Bioconjug Chem 8:23–30
5. Sakai H, Takeoka S, Wettstein R, et al (2002) Systemic and microvascular responses to hemorrhagic shock and resuscitation with Hb vesicles. Am J Physiol 283:H1191–H1199
6. Sakai H, Hara H, Yuasa M, et al (2000) Molecular dimensions of Hb-based O-2 carriers determine constriction of resistance arteries and hypertension. Am J Physiol 279: H908–H915
7. Wakabayashi Y, Takamiya R, Mizuki A, et al (1999) Carbon monoxide overproduced by heme oxygenase-1 causes a reduction of vascular resistance in perfused rat liver. Am J Physiol 277:G1088–G1096
8. Kyokane T, Norimizu S, Taniai H, et al (2001) Carbon monoxide from heme catabolism protects against hepatobiliary dysfunction in endotoxin-treated rat liver. Gastroenterology 120:1227–1240
9. Sakai H, Horinouchi H, Tomiyama K, et al (2000) Hemoglobin-vesicles as oxygen carriers: influence on phagocytic activity and histopathological changes in reticuloendothelial system. Am J Pathol 159:1079–1088
10. Agam G, Livne AA (1992) Erythrocytes with covalently bound fibrinogen as a cellular replacement for the treatment of thrombocytopenia. Eur J Clin Invest 22:105–112
11. Levi M, Friderich P, Ten CW, et al (1999) Fibrinogen-coated albumin microcapsules reduce bleeding in severely thrombocytopenic rabbits. Nat Med 5:107–111
12. Chao F, Reddick RL, Bode AR, et al (1996) Infusible platelet membrane microvesicles: a potential transfusion substitute for platelet. Transfusion 36:536–542
13. Soslau G, Class R, Morgan DA, et al (2001) Unique pathway of thrombin-induced platelet aggregation mediated by glycoprotein Ib. J Biol Chem 276:21173–21183
14. Nishiya T, Murata M, Handa M, et al (2000) Targeting of liposomes carrying recombinant fragments of platelet membrane glycoprotein Ibα to immobilized von Willebrand factor under flow conditions. Biochem Biophys Res Commun 270:755–760
15. Takeoka S, Teramura Y, Okamura Y, et al (2002) Rolling properties of rGPIbα-conjugated phospholipid vesicles with different membrane flexibilities on vWf surface under flow conditions. Biochem Biophys Res Commun 296:765–770
16. Kainoh M, Tanaka T (2002) Production of soluble integrin α2β1 heterodimer complex functionally in vitro and in vivo. Biochem Biophys Res Commun 290:305–310
17. Nishiya T, Kainoh M, Murata M, et al (2002) Reconstitution of adhesive properties of human platelets in liposomes carrying both recombinant glycoproteins Ia/IIa and Ibα under flow conditions: specific synergy of receptor-ligand interactions. Blood 100: 136–142
18. Takeoka S, Teramura Y, Ohkawa H, et al (2000) Conjugation of von Willebrand factor-binding domain of platelet glycoprotein Ib alpha to size-controlled albumin microspheres. Biomacromolecules 1:290–295
19. Teramura Y, Okamura Y, Takeoka S, et al (2003) Hemostatic effects of polymerized albumin particles bearing rGPIa/IIa in thrombocytopenic mice. Biochem Biophys Res Commun 306:256–260

Analysis of Angiogenic Profiles After Antiangiogenic Therapy by Colored Dye Extraction Microsphere Technique

Arifumi Iwamaru, Masazumi Watanabe, Takashi Ohtsuka, Hirohisa Horinouchi, and Kouichi Kobayashi

Key words. Angiogenic profile, Antiangiogenic therapy, Colored dye extraction microspheres

Introduction

A technique that can measure tumor blood flow easily, accurately and economically is required to study tumor angiogenesis and angiogenesis inhibition, because it can be used to obtain much information about antiangiogenic status within tumor before and after antiangiogenic therapy [1]. However, no study has directly estimated tumor blood flow after antiangiogenic therapy. Using dye extraction colored microspheres, we estimated blood flow in a rat ascites hepatoma LY80 tumor model before and after antiangiogenic therapy with TNP-470 [2,3].

Materials and Methods

An antiangiogenic agent, TNP-470 (30 mg/kg), was injected subcutaneously every other day from the day after transplantation of LY80 cells. Dyed microspheres were infused into the tumor-bearing rats via a catheter inserted to the left ventricle. After removal, the tumors were dissolved by a 4 M KOH solution to isolate the microspheres. The dye was extracted and its concentration was quantified by spectrophotometry. Tumor blood flow corrected for wet weight ($ml\,min^{-1}\,g^{-1}$) was calculated as follows (AU = absorbency units): [blood flow to tumor] = [AU per gram of tumor] × [reference withdrawal

Department of Surgery, Keio University School of Medicine, Shinanomachi 35, Shinjuku-ku, Tokyo 160-8582, Japan rate]/[AU per gram of reference blood]. Tumor tissues with or without TNP-470 were also examined histologically.

Results

Blood flow in the LY80 tumors negatively correlated with tumor weight. Tumor blood flow 1 week after transplantation could be predicted by the following simple regression equation: tumor blood flow ($ml\,min^{-1}g^{-1}$) = 0.95–0.09 × tumor weight (g). In the groups treated with TNP-470, this simple regression equation shifted to the left (Fig. 1), despite virtually no difference in the results of pathological examination. In LY80 tumors 2 weeks after transplantation, blood flow in tumors treated with TNP-470 was higher than without TNP-470 (Fig. 2).

Discussion

Our results suggest that TNP-470 has antiangiogenic activity and normalizes tumor blood flow 2 weeks after tumor transplantation. In addition, our colored microsphere technique was more useful than histological examination for detecting small changes within tumors after 1 week of antiangiogenic therapy.

In conclusion, this technique can be used to obtain angiogenic profiles of tumors before and after antiangiogenic therapy. Angiogenic profiles may

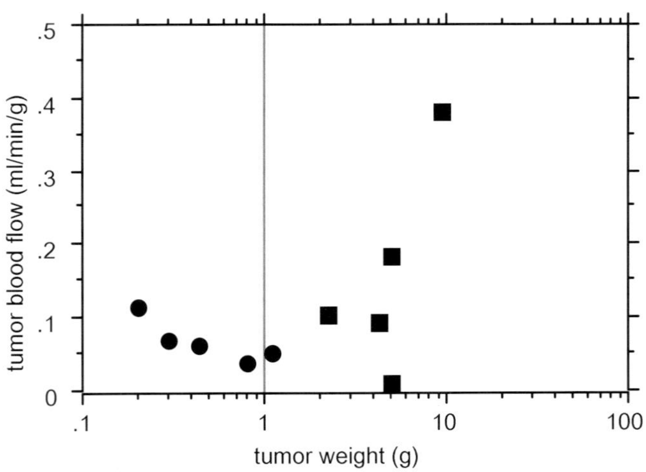

Fig. 1. In the TNP-470 treatment groups, blood flow in LY80 tumors at 2 weeks (*squares*) showed a wide range of values. Blood flow in LY80 tumors at 1 week (*circles*) negatively correlated with tumor weight

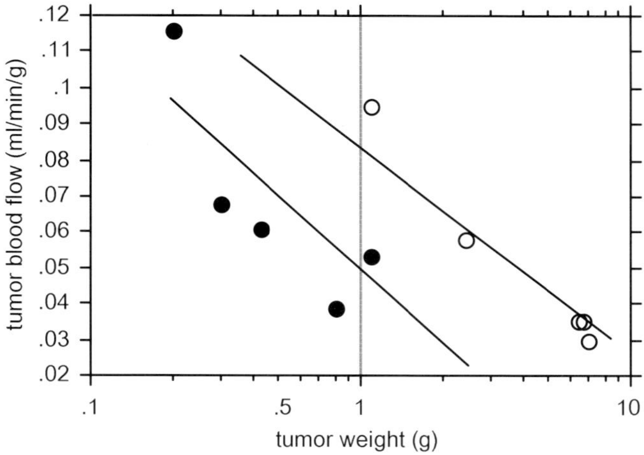

FIG. 2. LY80 tumor blood flow 1 week after transplantation in the TNP-470 treatment groups (*closed circles*) and LY80 tumor blood flow 1 week after transplantation (*open circles*). Simple regression analysis showed a shift of the *oblique line* to the left

provide important clues to optimal antiangiogenic therapy, combinations of antiangiogenic treatment and cytotoxic therapy, and schedules for combination therapy.

References

1. Jain RK (2001) Normalizing tumor vasculature with anti-angiogenic therapy: A new paradigm for combination therapy. Nat Med 9:987–989
2. Iwamaru A, Watanabe M, Yu S, et al (2001) Measurement of tumor blood flow using colored dye extraction microspheres in two rat tumor models. Int J Oncol 18:227–232
3. Iwamaru A, Watanabe M, Ohtsuka T, et al (2003) Analysis of angiogenic profiles by estimation of tumor blood flow with colored dye extraction microspheres after antiangiogenic therapy. Oncol Rep 10:1127–1131

Part 2
Gastroduodenal Microcirculation and Disease

Esophageal and Gastrointestinal Microcirculation: Essential for Mucosal Protection, a Target for Injury, and a Critical Component of Injury and Ulcer Healing

Andrzej S. Tarnawski, Jianyuan Chai, and Michael K. Jones

Summary. Blood flow through the network of mucosal microvessels in the gastrointestinal tract is essential for delivery of oxygen and nutrients to all mucosal constituents. The endothelial cells lining the microvessels, although protected by prostaglandins, are major targets for injury by various noxious factors, such as ethanol, nonsteroidal anti-inflammatory drugs (NSAIDs), ischemia–reperfusion, and free radicals. Endothelial injury results in formation of thrombi, microvascular stasis, and hypoxia, leading to tissue necrosis in the form of erosions or ulcers. Quantitative histologic, Transmission (TEM) and scanning electron microscopy (SEM) studies show that acute mucosal injury such as erosion triggers angiogenesis in the mucosal microvessels bordering necrosis: endothelial sprouting, formation of endothelial tubes ultimately leading to restoration of microvessels in regenerating tissue. The major molecular trigger for the initiation of angiogenesis in injured esophageal and gastric mucosa is accumulation of hypoxia inducible factor-1α (HIF-1α), which activates the genes encoding vascular endothelial growth factor (VEGF), its receptor, and angiopoietins that regulate angiogenesis. During healing of esophageal or gastric ulcers, granulation tissue, consisting of fibroblasts and proliferating endothelial cells forming microvessels, develops at the ulcer base. The newly formed microvessels, supported by fibroblasts and smooth muscle cells, sprout into the ulcerated area and restore the microvasculature in the ulcer scar. The molecular mechanisms stimulating angiogenesis in granulation tissue include activation of HIF-1α, egr-1, and genes encoding basic fibroblast growth factor, VEGF, angiopoietin-1, angiopoietin-2, their receptors, and cyclooxygenase-2. The latter co-localizes with upregulated VEGF in esophageal and gastric ulcers. The scars of healed ulcers demonstrate prominent microvascular abnormalities detected by cast/SEM

Department of Gastroenterology, VA Medical Center, Long Beach and the University of California, Irvine, 5901 East 7th Street, Long Beach, CA 90822-5201, USA and vascular permeability studies. NSAIDs inhibit angiogenesis by interfering with MAP/Erk2 kinase, actin cytoskeleton, and D_1 cyclin. Such actions are likely responsible for NSAID interference with ulcer healing. Our studies demonstrated that gene therapy with a single local injection of VEGF and angiopoietin-1 cDNAs stimulates angiogenesis, promotes restoration of microvascular network, and accelerates healing of experimental gastric and esophageal ulcers.

Key words. Endothelial cell, Hypoxia, Angiogenesis, Vascular endothelial growth factor, Gene therapy

Gastric Microcirculation

Blood flow through microvessels (capillaries, arterioles, and collecting venules) is crucial for maintaining the structure and functions of all tissues including esophageal and gastrointestinal mucosa [1–5]. The microcirculation is critical because it delivers oxygen and nutrients to all tissues and cells and removes toxic metabolites [3]. The vascular cast/scanning electron microscopy studies of normal gastric mucosal microvasculature demonstrated a very dense capillary network in the gastric mucosa in lamina propria adjacent to glandular epithelial cells (Fig. 1) [6].

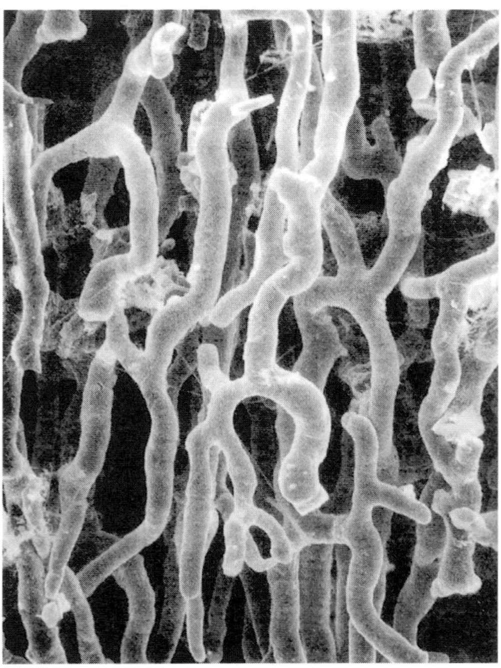

Fig. 1. Vascular capillary casts of normal gastric mucosa in rat. Vasculature was filled with Mercox resin (Dainippon, Tokyo, Japan). Thirty minutes later, the stomach was removed, cut into small pieces and immersed in 20% NaOH for 6h to dissolve tissue external to the cast. Desiccated specimens were coated with gold/palladium (60:40) and examined under a scanning electron microscope (5500, Hitachi, Tokyo, Japan) at 10kV. Reprinted, with permission, from our previous publication [6]

Microvascular Blood Flow, its Regulation, and its Response to Injurious Factors

The microvascular endothelium generates potent vasodilators such as nitric oxide (NO) and prostacyclin, which protect the gastric mucosa against injury and oppose the mucosal damaging action of vasoconstrictors such as leukotriene C_4, thromboxane A_2, and endothelin [3–5]. When the gastric mucosa is exposed to an irritant, or when acid back-diffusion occurs, there is, in response, a marked and rapid increase in mucosal blood flow [3–5], which allows removal and/or dilution of the back-diffusing HCl and/or noxious agents. This response is essential for mucosal defense, because its abolishment through mechanical restriction of blood flow leads to the development of hemorrhagic necrosis. Holzer [7] characterized the mechanisms responsible for this hyperemic response and demonstrated that it is mediated by sensory afferent nerves. Stimulation of gastric sensory nerves leads to the release of neurotransmitters such as calcitonin gene-related peptide (CGRP) and substance P in the nerve terminals, located within or close to the large submucosal vessels [7–9]. Calcitonin gene-related peptide exerts a mucosal protective action most likely through vasodilatation of submucosal vessels mediated by NO generation [9,10]. Ablation of the sensory afferent nerves (by chronic, large dose treatment with capsaicin) abolishes the hyperemic response and greatly increases the susceptibility of gastric mucosa to damage [4,7,9].

Nitric oxide, a potent vasodilator, is generated from L-arginine by an enzyme, NO synthase (NOS), which is expressed in endothelial cells as two isoforms: a constitutive (eNOS) and an inducible (iNOS) [3–5,9,11]. In the gastric mucosa, NO plays a major role in mucosal defense by regulating the amount of blood entering mucosal circulation. Endogenous and exogenous NO protects the gastric mucosa against injury by ethanol and endothelin-1, while inhibition of NOS (resulting in reduced NO generation) increases gastric mucosal injury [3,5].

Endothelial cells of microvessels are the major targets for injury by various noxious factors such as ethanol, nonsteroidal anti-inflammatory drugs (NSAIDs), ischemia–reperfusion, and free radicals. Gastric mucosa exposed to ulcerogenic and necrotizing agents such as aspirin, indomethacin, and other nonselective drugs (NSAIDs), bile acids, alcohol, ischemia, or corrosive agents develops characteristic morphologic, ultrastructural, and functional changes reflecting injury [3,4,11,12]. Damage to the microvascular endothelium leads to microvascular stasis, cessation of oxygen delivery, and transport of nutrients [3,4,8]. Microvascular damage occurs early during mucosal injury, precedes necrosis of glandular cells, and adds an ischemic component to the direct toxic injury of the cells [11].

Our endoscopic, histologic, and ultrastructural studies demonstrated that, after intragastric administration of ~50% ethanol (a concentration comparable to that in hard liquors), endothelial capillary injury occurs as early as 5 min (Fig. 2B) and leads to thrombi formation and microvascular stasis (Fig. 2C), resulting in ischemia and hypoxia, which in turn lead to local necrosis in the form of erosions [11]. Our study also demonstrated that in human gastric mucosa endothelial cells are also the major targets of prostaglandin-induced protection against alcohol-induced injury [12].

Arteriovenous (AV) shunting blood flow is another potential mechanism of mucosal injury. Kitajima's group demonstrated using intravital microscopy that AV blood shunting takes place in the gastric mucosa after thermal injury [13]. Such blood shunting can cause mucosal hypoperfusion resulting in hypoxia, which may contribute to erosion formation [13].

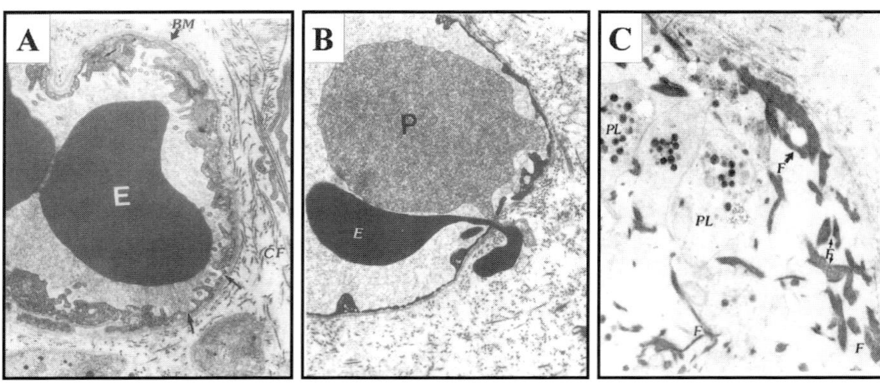

FIG. 2A–C. Transmission electron micrographs of human gastric mucosal microvessels. **A** The structure of normal capillary vessel in the mid mucosa. The capillary wall and endothelial cell cytoplasm is normal with a characteristic fenestration (*arrows*). *BM*, basement membrane; *E*, erythrocytes in the capillary lumen; *J*, junction between two neighboring endothelial cells; *CF*, collagen fibers (×17400). **B** Electron micrograph of human gastric mucosa 5 min after intragastric alcohol administration. The continuity of the capillary wall is broken and erythrocyte (*E*) together with coagulated plasma (*P*) are leaking into lamina propria (×17400). **C** Transmission electron micrograph of the gastric mucosa 15 min after alcohol administration. Part of a necrotic capillary wall is shown with aggregated platelets (*PL*) and deposited fibrin (*F*) (×21450). Reprinted, with permission, from our previous papers [11,12]

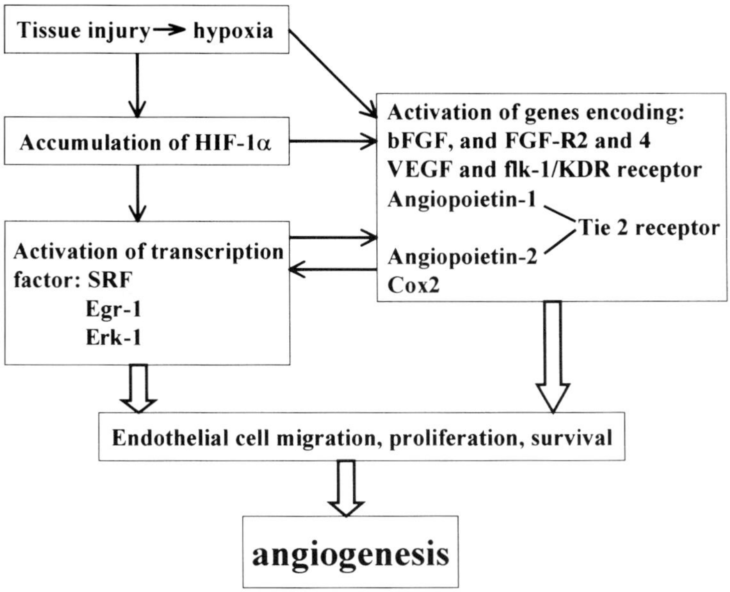

FIG. 3. Diagrammatic representation of molecular events leading to activation of angiogenesis. *HIF-1α*, hypoxia inducible factor-1α; *SRF*, serum response factor; *Erk*, extracellular regulated protein kinase; *Egr*, early growth response; *VEGF*, vascular endothelial growth factor; bFGF, basic fibroblast growth factor; *KDR*, kinase domain receptor; *Cox*, cyclooxygenase

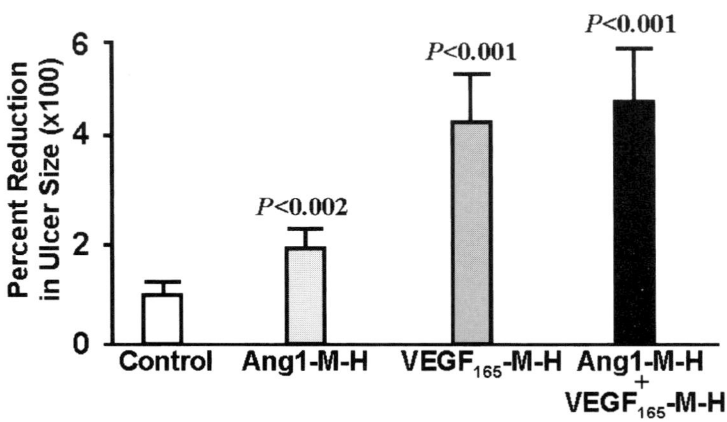

FIG. 4. Gene therapy with angiogenic growth factor cDNAs encoding vascular endothelial growth factor (*VEGF*) and angiopoietin-1 (*Ang1*) accelerates gastric ulcer healing: quantitative data for ulcer size reduction. Percent reduction in ulcer size was measured at 14 days after ulcer induction and injection of either control plasmid (set to 100%), Ang1–M-H, $VEGF_{165}$–M-H, or the combination of Ang1–M-H+$VEGF_{165}$–M-H ($n = 18$ for each group). Reprinted, with permission, from our previous paper [35]

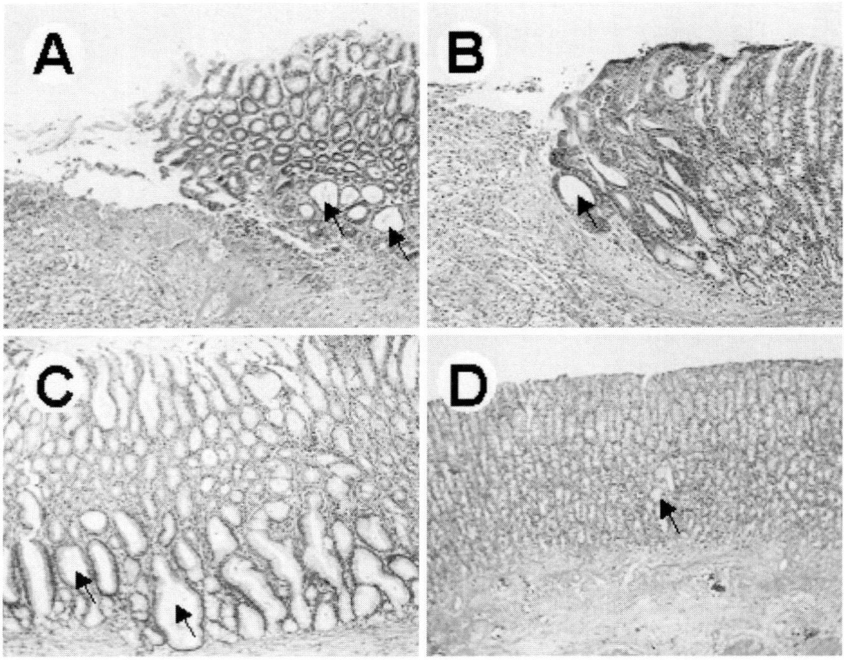

FIG. 5A–D. Gene therapy with VEGF and Ang1 improves mucosal healing assessed histologically. Histologic illustration of ulcer healing in representative groups at 14 days. Single injection of plasmids expressing $VEGF_{165}$ and Ang1 accelerates ulcer healing and scar formation. **A** In the control group (empty vector), a large ulcer is present with a moderate amount of granulation tissue. In the basal portion of the ulcer margin mucosa, several dilated glands are present. **B** Representative ulcer from the Ang1-treated group. Granulation tissue is more developed, compared to the control group, reaching about the height of the ulcer margin. Several tubes are budding into the granulation tissue (*arrow*). **C** In the VEGF-treated group 33% of the ulcers were completely healed. Mucosal scar contains several dilated gastric glands (*arrows*), but restoration of the mucosal architecture in the scar, even if imperfect, was accomplished. **D** In the VEGF+Ang1-treated group, 44% of the ulcers were completely healed. Restoration of the architecture is almost complete except for a few slightly dilated glands (*arrow*). × 100. Reprinted, with permission, from our previous paper [35]

Angiogenesis in Acutely Injured Mucosa and its Regulation

Angiogenesis, formation of new capillary blood vessels, is a fundamental process essential for embryonic development, tissue growth and reparative processes such as wound healing [14–17]. Under these conditions, the resting phenotype of endothelial cells is changed to an angiogenic phenotype [15].

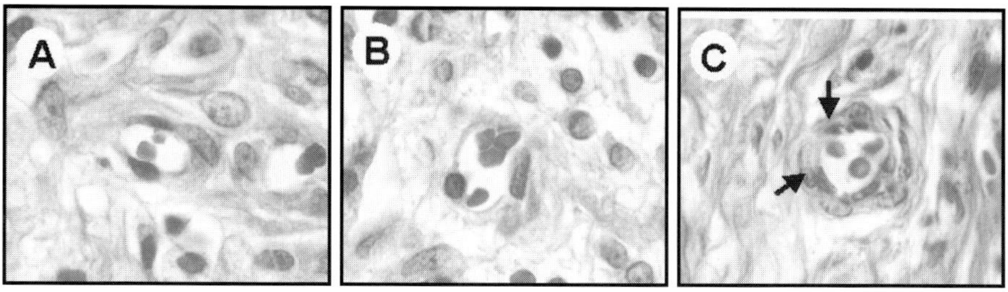

Fig. 6A–C. Gene therapy with VEGF plus Ang1 improves restoration of well-found and mature microvessels. Photomicrographs of granulation tissue stained by H&E. **A** Granulation tissue of ulcers injected with control plasmid; **B** plasmid-encoding $VEGF_{165}$; or **C** the combination of plasmids encoding both $VEGF_{165}$ and Ang1. The microvessels of the granulation tissue from the ulcers injected with either the control plasmid or the plasmid encoding $VEGF_{165}$ contained primarily only endothelial cells with very few pericytes. In contrast, the microvessels of the granulation tissue from the ulcers injected with the combined plasmids expressing both $VEGF_{165}$ and Ang1 appeared to be more well-formed and contained more pericytes (*arrows*). Pericytes were identified by their close adherence to the endothelium and characteristic shape of being flat toward the abluminal surface of the endothelium and convex toward the surrounding collagen. $\times$ 400. Reprinted, with permission, from our previous paper [35]

As a result, the microvascular endothelial cells from preserved microvessels at the wound edge migrate, proliferate and initiate re-establishment of a microvascular network through the process of angiogenesis [14,17]. Our studies demonstrated that 24 h after alcohol injury, at the edge of non-necrotic gastric mucosa, about 6%–8% of mucosal microvessels undergo angiogenesis: basement membrane dissolution, endothelial cell budding into the extravascular space, tube formation and, ultimately, reconstruction of a capillary network [14]. Angiogenesis in injured gastric mucosa is associated with a significant increase in the expression of Ki-ras mRNA and Ras protein and a significant 4–6-fold increase in vascular endothelial growth factor (VEGF) mRNA and protein levels that trigger gastric mucosal angiogenesis in response to ethanol injury [14]. The gastric mucosal angiogenesis is strongly stimulated by VEGF, basic fibroblast growth factor (bFGF), prostaglandin E_2, and prostacyclin [18].

Ulcer Formation: Importance of Vascular Lesions

Experimental studies have clearly demonstrated that vascular and microvascular changes are the earliest events in, as well as the basis for, the development of experimental gastric ulcers [19]. These vascular changes (thrombi, constriction) cause mucosal ischemia, free radical formation, and cessation of nutrient delivery, all resulting in ischemic necrosis of the mucosa and muscularis mucosae [19].

Ulcer Healing: Reconstruction of the Microvessels Through Angiogenesis in Granulation Tissue

Granulation tissue develops at the ulcer base within 48–72 h after ulceration [19]. It consists of proliferating connective tissue cells, namely, macrophages, fibroblasts, and proliferating endothelial cells, which form microvessels through the process of angiogenesis [19]. Granulation tissue is important for the ulcer-healing process because it supplies microvessels and connective tissue cells for restoration of the microvasculature within the ulcer scar [19]. The growth of granulation tissue and generation of new microvessels through angiogenesis is stimulated by bFGF, VEGF, platelet-derived growth factor (PDGF), angiopoietins (Ang1 and Ang2), ephrins, leptin, cyr61, prostaglandins, and cytokines, including interleukin-1 and tumor necrosis factor-α [18,20].

Vascular endothelial growth factor is a potent angiogenic growth factor and is an endothelial cell-specific mitogen because its receptors, flt-1 and flk-1, are restricted predominantly to endothelial cells, although other cells such as endothelial cell precursors and some cancer cells also express these receptors (reviewed in [18,20]). Vascular endothelial growth factor (originally identified as a vascular permeability factor) is produced by vascular smooth muscle cells, tumor cells, and endothelial cells [20], and has been implicated in stimulation of the normal physiological angiogenesis involved in wound and ulcer healing [14,16,20]. Activation of the flk-1 receptor tyrosine kinase in response to VEGF is necessary for VEGF-induced angiogenic sprouting, which is facilitated by activation of matrix metalloproteinases that degrade the extracellular matrix [20]. Vascular endothelial growth factor, bFGF, and Ang1 stimulate endothelial cell proliferation and migration that result in microvascular network assembly [15,16,20]. Vascular endothelial growth factor production is stimulated by PDGF, transforming growth factor $\beta 1$, and bFGF, as well as by cytokines, NO, E-series prostaglandins (PGs), and hypoxia [18,20]. Hypoxia stimulates VEGF production via the hypoxia-inducible transcription factor (HIF)-1, which accumulates in response to hypoxia and activates transcription from a hypoxia response element (HRE) binding site in the VEGF promoter [18–20]. Vascular endothelial growth factor is also regulated post-transcriptionally under hypoxia [20–27]. For example, it has been shown that VEGF mRNA is stabilized under hypoxia by binding of the Elav-like protein, HuR, to specific sequences within the 3′-untranslated region, thereby increasing the VEGF mRNA half-life > 2-fold and compounding the increases in VEGF mRNA levels produced by HIF-induced transcription under hypoxia [27].

The angiogenic peptides Ang1 and Ang2, and the endothelial specific receptors Tie1 and Tie2, are involved in angiogenic processes occurring subsequent to the actions of VEGF and its receptors [14,15,22,29–31]. It is known that Ang1 acts to stabilize vessels and promotes vessel maturation and, therefore, Ang1 *does* function subsequent to the actions of VEGF. However, Ang2 functions to inhibit the Ang1–Tie2 interaction, thereby destabilizing vessels and allowing VEGF access to receptors on endothelial cells. Ang2 probably does not function subsequent to VEGF but rather promotes VEGF function and attracts recruitment of vascular smooth muscle cells [30,31].

We have demonstrated that esophageal ulceration induces HIF-1α protein expression and VEGF gene activation reflected by increased VEGF mRNA (240%) and VEGF protein (310%) levels. HIF-1α protein was expressed in microvessels bordering necrosis where it co-localized with VEGF [28].

NSAIDs Interfere with Angiogenesis

Endothelial cells of the microvessels are also major targets of NSAID-induced injury to gastrointestinal mucosa [32]. Moreover, NSAIDs inhibit angiogenesis (reviewed in [18,33,34]). Based on the experimental data and the literature, the mechanisms by which NSAIDs inhibit angiogenesis appear to be multifactorial and likely include local changes in angiogenic growth factor expression, alteration in key regulators and mediators of VEG, increased endothelial cell apoptosis, inhibition of endothelial cell migration, recruitment of inflammatory cells and platelets, and/or thromboxane A_2-mediated effects (reviewed in [18]). Some of these mechanisms include: inhibition of mitogen-activated, or extracellular regulated protein kinase (Erk) activity; suppression of cell cycle proteins; inhibition of early growth response (Egr-1) gene activation; interference with HIF-1 and VEGF gene activation; increased production of the angiogenesis inhibitor, endostatin; inhibition of endothelial cell proliferation, migration, and spreading; and the induction of endothelial apoptosis (reviewed in [18,34]).

Gene Therapy with Angiogenic Growth Factors

We have previously reported that endogenous VEGF is required for unimpeded healing of mucosal injury [13]. As a follow-up of this work, we performed studies to determine whether a single local injection of VEGF and Ang1 cDNA (gene therapy), with limited duration of target gene expression, affects healing of experimental gastric ulcers [35]. This study demonstrated that a single injection of plasmid DNA-encoding VEGF increases angiogenesis in granulation tissue by 7 days after injection; and that this increase strongly correlates with a reduction in ulcer size by more than 400% at 14 days compared with controls injected with nonexpressing plasmid [35]. Simultaneous injection of both the VEGF and Ang1 expression plasmids did not significantly decrease ulcer size beyond that resulting from injection of the VEGF expression plasmid alone (possibly indicating that the angiogenic response to VEGF has reached a maximum). However, with the combined treatment of VEGF+Ang1, the microvessels in the granulation tissue contained more pericytes, were better developed, and the structural restoration of gastric mucosa within the scar was more complete [35]. A neutralizing anti-VEGF antibody abolished VEGF gene therapy-induced accelerated ulcer healing, indicating an essential role for VEGF and enhanced angiogenesis in ulcer healing [35]. Gene therapy with cDNA encoding VEGF also significantly enhances angiogenesis and accelerates healing of experimental esophageal ulcers, indicating essential roles of VEGF and angiogenesis in esophageal ulcer healing [28].

Pharmacological Stimulation of Angiogenesis for Gastric Mucosal Injury and Ulcer Healing: Present and the Future

Experimental studies demonstrated that therapy with bFGF, PDGF, and/or VEGF accelerates angiogenesis and accelerates healing of duodenal ulcers in rats [36,37]. Our studies (described above) have demonstrated efficacy of gene therapy with angiogenic growth factors for the healing of experimental gastric and esophageal ulcers [28,35]. However, application of the above therapies for human gastrointestinal ulcers will require at least 5–7 years. While application of angiogenic growth factors and gene therapy for angiogenesis-based healing of human gastrointestinal ulcers still requires clinical trials, a currently available drug, rebamipide, has been shown to activate genes encoding angiogenic growth factors (e.g. VEGF > 7-fold, HB-EGF > 5-fold), and cyclooxygenase > 9.3-fold in gastric mucosal cells and, in addition, to directly stimulate angiogenesis [38,39].

Acknowledgments. This study was supported by the Medical Research Service of the Department of Veterans Affairs, Merit Reviews (A.S.T. and M.K.J) and Research Enhancement Award Program (A.S.T.).

References

1. Guth PH, Leung FW (1987) Physiology of the gastric circulation. In: Johnson LR (ed) Physiology of the gastrointestinal tract. Raven, New York, pp 1031–1054

2. Crissinger KD, Granger N (1995) Gastrointestinal blood flow. In: Yamada T (ed) Textbook of gastroenterology, 2nd edn. Lippincott, Philadelphia, pp 518–545
3. Tarnawski A (1998) Cellular and molecular mechanisms of mucosal defense and repair. In: Bioregulation and its disorders in gastrointestinal tract. Blackwell Science, Tokyo, pp 3–17
4. Tarnawski A, Jones MK, Baatar D, et al. (2002) Role of angiogenesis and angiogenic growth factor in mucosal repair and ulcer healing. In: Cho CH, Wang JY, Wang JH (eds) Karger gastrointestinal mucosal repair and experimental therapeutics, vol. 25. pp 101–116
5. Kawano S, Tsuji S (2000) Role of mucosal blood flow: a conceptual review in gastric mucosal injury and protection. J Gastroenterol Hepatol 15:D1–D6
6. Ichikawa Y, Tarnawski A, Sarfeh IJ, et al (1994) Distorted microangio-architecture and impaired angiogenesis in gastric mucosa of portal hypertensive rat. Gastroenterology, 106:706–708
7. Holzer P (1991) Sensory nerves and neuropeptides in gastroenterology. From basic science to clinical perspectives. Adv Exp Med Biol 298:3–17
8. Tarnawski A, Sarfeh IJ, Lu S-Y, et al (1993) Quality of ulcer healing: evidence for impaired restoration of sensory CGRP nerves in the scar of experimental gastric ulcer. Eur J Gastroenterol Hepatol 5(3):S81–S85
9. Peskar BM, Plate S, Stroff T (1997) Gastroprotective effects of gut peptides: Role of afferent neurons, calcitonin gene-related peptide and nitric oxide. In: Cheli R, Iaquinto G, Szabo S (eds) Gastroduodenal mucosal damage: problems of protection and healing. Medserve, Milan, pp 70–83
10. Arai K, Ohno T, Saeki T, et al (2003) Endogenous prostaglandin 12 regulates the neural emergency system through release of calcitonin gene related peptide. Gut 52(9): 1242–1249
11. Tarnawski A, Stachura J, Gergely H, et al (1988) Microvascular endothelium: A major target for alcohol injury of the human gastric mucosa: histochemical and ultrastructural study. J Clin Gastroenterol 10(suppl 1):S53
12. Tarnawski A, Stachura J, Hollander D, et al (1988) Cellular aspects of alcohol-induced injury and prostaglandin protection of the human gastric mucosa. J Clin Gastroenterol 10(suppl 1):S35–S45
13. Yoshida M, Wakabayashi G, Ishikawa H, et al (2002) Arteriovenous shunting blood flow is intravitally observed in the stomach after thermal injury in rats. Keio J Med 51(4):193–200
14. Jones MK, Itani RM, Wang H, et al (1999) Activation of VEGF and Ras genes in gastric mucosa during angiogenic response to ethanol injury. Am J Physiol 276:G1345–G1355
15. Folkman J, D'Amore PA (1996) Blood vessel formation: what is its molecular basis? Cell 87:1153–1155
16. Carmeliet P (2003) Angiogenesis in health and disease. Nat Med 9:653–693
17. Tarnawski A, Hollander D, Stachura J, et al (1991) Role of angiogenesis in healing of experimental gastric ulcer. In: Halter F, Garnet A, Tytat GNJ (eds) Mechanisms of peptic ulcer healing. Kluwer, Dordrecht, pp 165–171
18. Tarnawski AS, Jones MK (2003) Inhibition of angiogenesis by NSAIDs: molecular mechanisms and clinical implications. J Mol Med 81:627–636
19. Tarnawski A, Hollander D, Stachura J, et al (1999) Vascular and microvascular changes—key factors in the development of acetic acid-induced gastric ulcers in rats. J Clin Gastroenterol 22(suppl 1):S148–S157

20. Ferrara N, Gerber H-P, LeCourter J (2003) The biology of VEGF and its receptor. Nat Med 9:669–676
21. Pugh CW, Ratcliffe PJ (2003) Regulation of angiogenesis by hypoxia: role of the HIF system. Nat Med 9:677–684
22. Risau W (1997) Mechanisms of angiogenesis. Nature 386:671–673
23. Levy AP, Levy NS, Goldberg MA (1996) Post-transcriptional regulation of vascular endothelial cell growth factor by hypoxia. J Biol Chem 271:2746–2753
24. Gerber HJP, Condorelli F, Park J, Ferrara N (1997) Differential transcriptional regulation of the two vascular endothelial growth factor receptor genes: Flt-1, but not Flk-1/KDR, is up-regulated by hypoxia. J Biol Chem 272:23659–23667
25. Kelly BD, Hackett SF, Hirota K, et al (2003) Cell type-specific regulation of angiogenic growth factor gene expression and induction of angiogenesis in nonischemic tissue by a constitutively active form of hypoxia-inducible factor-1. Circ Res 93(11):1074–1081
26. Yamakawa M, Liu LX, Date T, et al (2003) Hypoxia-inducible factor-1 mediates activation of cultured vascular endothelial cells by inducing multiple angiogenic factors. Circ Res 93(7):664–673
27. Levy NS, Chung S, Furneaux H, et al (1998) Hypoxic stabilization of vascular endothelial growth factor mRNA by the RNA-binding protein HuR. J Biol Chem 13; 273(11):6417–6423
28. Baatar D, Jones MK, Tasugawa K et al (2002) Esophageal ulceration triggers expression of hypoxia-inducible factor-1α and activates vascular endothelial growth factor gene. Am J Pathol 161:1449–1457
29. Davis S, Aldrich TH, Jones PF, et al (1996) Isolation of angiopoietin-1, a ligand for the TIE2 receptor, by secretion-trap expression cloning. Cell 87:1161–1169
30. Maisonpierre PC, Suri C, Jones PF, et al (1997) Angiopoietin-2, a natural antagonist for Tie2 that disrupts in vivo angiogenesis. Science 277:55–60
31. Iivanainen E, Nelimarkka L, Elenius V, et al (2003) Angiopoietin-regulated recruitment of vascular smooth muscle cells by endothelial-derived heparin binding EGF-like growth factor. FASEB J 17(12):1609–1621
32. Tarnawski A, Stachura J, Gregeley H, et al (1990) Gastric microvascular endothelium: a major target for aspirin-induced injury and arachidonic acid protection. An ultrastructural analysis in the rat. Eur J Clin Invest 20(4):432–440
33. Tarnawski A, Stachura J, Douglass TG, et al (1991) Indomethacin impairs quality of experimental gastric ulcer healing: A quality histologic and ultrastructural analysis. In: Garner A, O'Brien PE (eds) Mechanisms of injury, protection and repair of the upper gastrointestinal tract. Wiley, Chichester, pp 521–531
34. Jones MK, Wang H, Levin E, et al (1999) Inhibition of angiogenesis by NSAIDs. Insight into the mechanisms and implications for cancer growth and ulcer healing. Nat Med 5:1418–1423
35. Jones MK, Kawanaka H, Baatar D, et al (2001) Gene therapy for gastric ulcers. Single local injection of VEGF and angiopoietin-1 DNAs dramatically accelerates gastric ulcers healing and improves quality of scar. Gastroenterology 121:1040–1047
36. Folkman J, Szabo S, Stovroff M, et al (1991) Duodenal ulcer. Discovery of a new mechanisms and development of angiogenic therapy that accelerates healing. Ann Surg 214:414–427
37. Szabo S, Folkman J, Vincze A, et al (1997) Modulation of vascular factors by VEGF/VPF is sufficient for chronic ulcer healing and acute gastroprotection (abstract). Gastroenterology 122:A303

38. Tarnawski A, Chai J, Chiou S-K, et al (2003) Rebamipide—a novel ulcer healing drug—strongly stimulates angiogenesis by activating genes inducing angiogenic growth factors, and Cox2. A key to its ulcer healing action? In: Pasteur Institute Euroconference Paris, Proceedings A-38
39. Tarnawski AS, Chai J, Pai R, et al (2004) Rebamipide activates genes encoding angiogenic growth factors and Cox2 and stimulates angiogenesis: a key to its ulcer healing action? Dig Dis Sci 49(2):202–209
40. Yoshida M, Wakabayashi G, Ishikawa H, et al (2002) Arteriovenous shunting blood flow is intravitally observed in the stomach after thermal injury in rats. Keio J Med 51(4):193–200

Microvascular Pathophysiology in Gastric Mucosal Inflammation Associated with *Helicobacter pylori* Infection

Hidekazu Suzuki[1], Tatsuhiro Masaoka[1], Masayuki Suzuki[2], and Hiromasa Ishii[1]

Summary. In the multitude of mechanisms involved in the development of gastric mucosal inflammation, derangement of the microcirculatory system is a common initial pathway. In response to gastric luminal antigens such as bacteria and food factors, the microvascular endothelium transduces signals to circulating cells. *Helicobacter pylori* colonization induces a significant level of leukocyte recruitment to the gastric mucosa as a result of the sequential pathological processes in the microcirculatory system. This bacterial infection is associated with a variety of clinical outcomes, including benign and malignant gastroduodenal diseases. After *H. pylori* inoculation, the bacteria colonize the surface epithelium, upon which numerous leukocytes rolling or adhering to the endothelial cells can be observed in the venules of the gastric mucosa. The adherent leukocytes migrate into the gastric mucosal parenchyma. Among the numerous proinflammatory factors, such as cytokines, free radicals, and enzymes, a monochloramine, derived from both leukocytes and *H. pylori*, is an important factor involved in the development of *H. pylori*-associated disease. Persistent *H. pylori* infection has recently been achieved in Mongolian gerbil models, in which the sequential microcirculatory and histopathological changes in the gastric mucosa have been shown to closely mimic the changes leading to gastric mucosal lesion formation in association with *H. pylori* infection in humans. Gastric mucosa infected with *H. pylori* exhibited significantly higher gastritis scores. Because the recently isolated growth-hormone-releasing peptide ghrelin is secreted mainly from the A-like cells in the gastric corpus, the anatomical extension of *H. pylori*-associated inflammation to the gastric corpus mucosa might also affect ghrelin secretion and its associated appetite control.

[1] Department of Internal Medicine, Keio University School of Medicine, 35 Shinanomachi, Shinjuku-ku, Tokyo 160-8582, Japan
[2] Department of Gastroenterology, National Tokyo Medical Center, 2-5-1 Higashigaoka, Meguro-ku, Tokyo 152-8902, Japan

Key words. H. pylori, Inflammation, Irritation, Ghrelin, Ischemia-reperfusion, Monochloramine

Introduction

Forty-six years ago, the late Professor Masasharu Tsuchiya (Fig. 1), our mentor, visited the laboratory of Dr. James Reilly at the Claude Bernard Hospital in Paris. Dr. Reilly was an experimental pathologist who proposed the concept of the irritation syndrome, as follows: When animals receive either supraphysiologic stimulation or relatively weak but repeated stimulation (irritation) at any particular site of the autonomic nervous system, the capillary vessels in that region exhibit transient constriction followed by dilatation, finally resulting in hemorrhagic lesion formation [1]. The hemorrhagic lesions could occur not only in the organ receiving the stimulation, but also in distal organs not exposed directly to the stressors [1]. A similar concept of stress responses through the pituitary-adrenal system has been widely known over the world as the "general adaptation syndrome," which was described by Dr. Hans Selye.

When Dr. Tsuchiya left Paris, Dr. Reilly introduced him to a famous and provocative article by Dr. Benjamin W. Zweifach [2] which described the concept of microcirculation, and recommended that Dr. Tsuchiya start microvascular research in Japan. The concept of microcirculation established by another of my mentors, the late Professor Benjamin W. Zweifach, proposed that the microcirculatory system contains multi-level functional microvessels, such as the arterioles, capillaries and venules [2]. Among the multistep events in nonspecific microvascular responses, platelet activation might be the initial event involved in the recruitment of white blood cells, contributing to the development of inflammatory lesions. The world-famous microcirculation

FIG. 1. Masaharu Tsuchiya (1928—2001). Late Professor Emeritus of Keio University, Former President of the Japanese Society for Microcirculation

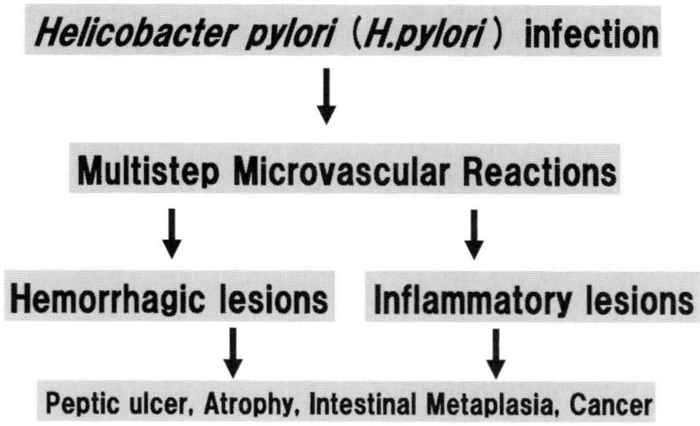

FIG. 2. Microvascular pathophysiology associated with *Helicobacter pylori* infection

laboratory in the University California, San Diego (UCSD), was taken over by Professor Geert W. Schmid-Schönbein, my mentor. I (Hidekazu Suzuki) spent 31 months with Dr. Zweifach, Dr. Schmid-Schönbein, and Mr. Frank A. DeLano in the microcirculation laboratory, Department of Bioengineering, UCSD, and examined several microvascular sequelae of ischemia–reperfusion and high blood pressure during my stint there [3–12].

In 1987, when Dr. Tsuchiya organized the 4th World Congress for Microcirculation in Tokyo, the concept of "Organ microcirculation" was introduced. Dr. Tsuchiya used to say to us, "Think about the whole system, and not too much about molecules in isolation". We really miss him.

Helicobacter pylori

Helicobacter pylori, a microaerobic Gram-negative rod, has flagellae and swims in the gastric mucus layer [13]. This bacterium specifically colonizes the gastric mucosa and possesses urease which neutralizes gastric acid to release ammonia (NH_3). In the gastric mucus, *H. pylori*, which possesses flagellae, exhibits corkscrew movements and penetrates onto the surface epithelium of the stomach [14,15]. The bacterial movements can be clearly detected by dark-field microscopy [16]. By tracing each coordinate, the velocity of each bacterium can be computed. This small organism could alter the microvascular responses in gastric mucosa and affect the gastric lesion formation such as peptic ulcer disease, gastric cancer, etc. (Fig. 2).

Helicobacter pylori and Microvascular Endothelial Cells

Kalia et al. [17] demonstrated in vitro that *H. pylori* strains exert cytostatic but not cytotoxic effects on the microvascular endothelial cells, suggesting that the microcirculatory disturbances observed in vivo may not be the result of direct endothelial cell damage. They also suggested that inhibition of angiogenesis might explain why ulcer healing is delayed in *H. pylori*-infected patients [17], and recently demonstrated that *H. pylori* suppressed the proliferative stage of angiogenesis in vitro [18]. On the other hand, Elizalde et al. [19] reported that platelet activation and aggregation contributed to the microvascular dysfunction and inflammatory cell recruitment associated with *H. pylori* infection in mice and humans.

Leukocytes and *H. pylori*

Activated neutrophils migrate from the microvasculature to the gastric mucosa in response to *H. pylori*-induced release of chemokines such as interleukin (IL)-8 and GRO [20]. *Helicobacter. pylori*-derived NH_3 also plays an important role in the pathogenesis of *H. pylori*-associated inflammation [21].

During the activation process of leukocytes, oxygen reacts with nicotinamide adenine dinucleotide phosphate (NADPH) oxidase of leukocytes, to yield superoxide (O_2^-). Then O_2^- reacts with superoxide dismutase (SOD) to yield hydrogen peroxide (H_2O_2). Neutrophils possess myeloperoxidase (MPO), which catalyzes the oxidation of chloride by H_2O_2 to yield hypochlorous anions (OCl^-). The interaction between *H. pylori*-derived NH_3 and OCl^- yields monochloramine (NH_2Cl), which is an exceptionally reactive and toxic oxidant, since it has a low molecular weight and strong lipophilic properties [21–23].

Helicobacter pylori in conjunction with the activated neutrophils exerts cytotoxic effects in vitro. Catalase, taurine, or methionine (NH_2Cl scavenger), and acetohydroxamic acid (urease inhibitor) were significantly effective at attenuating these injuries, suggesting the specific role of NH_2Cl in such injury [21]. On the other hand, apoptosis, a physiological suicide mechanism of cells that has a role in homeostasis, occurs during tissue turnover. NH_2Cl has also been shown to induce gastric epithelial apoptosis [22,23].

Helicobacter pylori lipopolysaccharide can not only induce the expression of iNOS in the stomach and duodenum following systemic challenge, resulting in gastroduodenal microvascular dysfunction [24], but can also provoke iNOS-mediated acute systemic microvascular inflammatory responses in the rat cardiac, hepatic, renal and pulmonary tissues [25].

Under the systemic irritable stimuli, ischemia–reperfusion injury is observed in gastric mucosal microcirculation. Such a kind of microvascular flow instability evokes oxidant formation in endothelial cells through xanthine oxidase or in the infiltrated leukocytes through NADPH oxidase. Generated oxidants will form NH_2Cl through the reaction with *H. pylori*-derived NH_3, leading to the epithelial injury including apoptosis (Fig. 3).

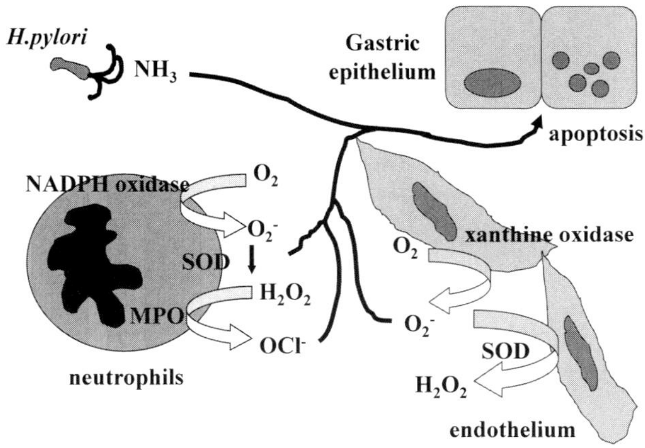

FIG. 3. Schematic mechanism of gastric mucosal oxidative stress under ischemia—reperfusion with *H. pylori* infection. *MPO*, myeloperoxidase; *SOD*, superoxide dismutase; *NADPH*, nicotinamide adenine dinucleotide phosphate

Animal Model

A recently introduced animal model of *H. pylori*-associated gastric disease, established by inoculation of this bacterium into Mongolian gerbils, has been shown to be useful for in vivo evaluation. Previously, we succeeded in visualizing the gastric microvascular responses to *H. pylori* in Mongolian gerbils using an intravital videomicroscope [14]. Leukocyte rolling and adhesion were observed on the venular endothelium in vivo in response to *H. pylori* challenge. The velocity of the leukocyte rolling was analyzed by the playback image analysis of the recorded video images. There was an increase in the tissue MPO activity as well as the content of thiobarbituric acid-reactive substances (marker of lipid peroxidation) in the *H. pylori*-infected gastric mucosa [26].

In the normal uninfected gastric mucosa of Mongolian gerbils, a capillary network is seen between the large collecting venules. The epithelial cell layer that covers the surface of the gastric mucosal layer exhibits a paved-stone-like appearance. The pores of the gastric glands that secrete acid and pepsin open onto the surface throughout the extent of the gastric mucosa.

In the venules of the *H. pylori*-colonized gastric mucosa of Mongolian gerbils, as time progressed the leukocyte rolling velocity decreased, increasing the chances of leukocyte–endothelial cell interactions, thereby promoting leukocyte adhesion and migration. In the initial stages of *H. pylori* infection, such as within 2 weeks after the inoculation, no significant changes of MPO activity were observed; however, a reduction in the leukocyte rolling velocity and increase in the extent of leukocyte adhesion were observed microscopically. The zinc-chelate compound, polaprezinc, significantly attenuated these microvascular changes in response to *H. pylori* [14].

In the multiple steps to gastric ulcerogenesis, opening up of arteriovenous (A-V) channels is one of the possible mechanisms responsible for the reduced mucosal blood flow as was previously reported by Professor Kitajima. Yoshida et al. intravitally observed the blood flow through A-V shunts in the stomach after thermal injury in rats [27]; they suggested, based on their observations, that A-V shunting could be one of the causes of mucosal hypoperfusion.

Ghrelin and *H. pylori* Infection

Ghrelin was isolated in 1999 from the human and rat stomach as an endogenous ligand of the growth hormone secretagogue receptor by Dr. Kangawa's group. Ghrelin is a 28-amino-acid peptide and possesses an *n*-octanoyl modification at the third serine residue, which has been shown to be necessary for its physiological activity [28]. It is produced and secreted from a subset of endocrine cells named A-like cells within the oxyntic glands of the gastric fundus [29]. The physiological roles of ghrelin are stimulation of growth hormone release, gastric motility and food intake [30]. The rat ghrelin sequence was found to correspond to a 28-residue signal sequence which has been also elucidated in humans, mice and Mongolian gerbils [31]. Studies on the structure–activity relationship of the peptide showed that the octanoyl group on the third serine is an essential moiety for this peptide to exert biological activity.

Recently, Nwokolo et al. [32] demonstrated that following eradication of *H. pylori* from asymptomatic patients, the plasma ghrelin levels increased profoundly, raising the possibility that ghrelin suppression by *H. pylori* might contribute to thin stature, and prevent the development of obesity-associated diseases. However, the situation in Mongolian gerbils was not quite so simple [31].

According to our data from animal experiments [31], chronically infected gerbils exhibited reduced gastric ghrelin concentrations, but because the inflamed stomach was enlarged, the total ghrelin content of the stomach remained unaffected. The weight of the gerbils remained unchanged, and unexpectedly, the plasma ghrelin levels were slightly increased. Histologically, the number of ghrelin-immunoreactive cells, located in the mucosal neck portion of the gastric fundus, especially in the vicinity of H^+-, K^+-ATPase-positive parietal cells, gradually decreased with an increase in the number of inflammatory cells in the gastric mucosa after *H. pylori* colonization. With progression to severe atrophy, extensive depletion of ghrelin-immunoreactive cells was noted [31]. It remains to be resolved whether *H.*

pylori infection also affects the ghrelin dynamics over the longer term, and whether or not suppressed ghrelin is important for the prevention of obesity and its obesity-associated diseases in patients with *H. pylori* infection.

Ischemia–reperfusion injury is also one of the important pathological processes in gastric mucosal inflammation. Gastric ischemia was induced for 30 min by applying a small vascular clamp to the celiac artery and reperfusion was produced by removal of the clamp in male Wistar rats treated under the food deprivation. The area of gastric mucosal erosion increased significantly after 60 min of reperfusion (Fig. 4). The level of rat plasma ghrelin tends to increase after the ischemia–reperfusion.

Helicobacter pylori and Extragastric Microvascular Diseases

A possible association between *H. pylori* infection and the development of coronary heart disease, thrombo-occlusive cerebral disease, or both, in diabetic patients has been reported [33]. Quadri et al. [34] reported that a higher prevalence of *H. pylori* infection was found in type-2 diabetic women than in control women, whereas there was no difference among diabetic and nondiabetic males. According to their reports [34], however, microangiopathy was significantly more prevalent in *H. pylori*-negative than in *H. pylori*-positive patients. Their assessment suggests that microvascular changes in the gastric mucosa may create an unfavorable environment for the establishment or survival of *H. pylori*.

Warzecha et al. reported that *H. pylori* infection could increase the severity of ischemia-induced pancreatitis [35]. They also demonstrated in rats that *H. pylori* infection could increase the production of pro-inflammatory IL-1β, and aggravate disturbances of the pancreatic microcirculation in acute pancreatitis [35]. In conclusion, *H. pylori* modifies microvascular responses not only in the gastric mucosa, but also in extragastric sites, even though this bacterium only colonizes the gastric pyloric mucosa.

Acknowledgments. This study was supported by a Grant-in-Aid for Scientific Research C (2) from the Japan Society for the Promotion of Science (JSPS) (15590686, to H.S.), and a grant from Keio University. The authors thank Dr. Sachiko Nomura, Miss Yuriko Minegishi, and Miss Kumiko Kurabayashi, Laboratory of GI Ecology & Antibody-Food, Center for Integrated Medical Research, Keio University School of Medicine, for their technical assistance. As the Secretary General of The 14th Keio University International Sympo-

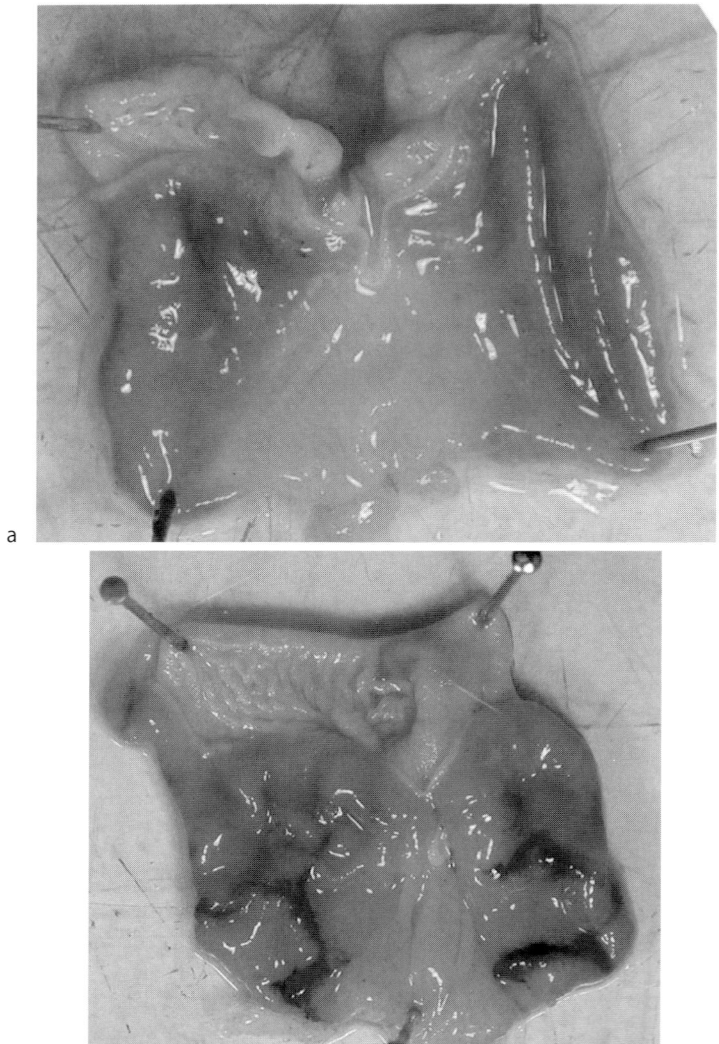

Fig. 4. Representative macroscopic findings of the rats without (**a**) and with (**b**) ischemia—reperfusion (30 min ischemia and 60 min reperfusion)

sium for Life Sciences and Medicine "Organ Microcirculation: A Gateway to Diagnostic and Therapeutic Interventions," I thank all the participants, including our guests from abroad. This symposium was successfully conducted, with some very useful discussions, thanks to their cooperation. We hope that the symposium will contribute significantly to progress in the field of Organ Microcirculation.

References

1. Reilly J, Rivalier E, Compagnon A (1934) Hemorrhagie, lesions vasculares et lymphatiques due tube digestif determinees par l'injection perisplanchnique de substances diverses. Compt Rend Soc Biol 116:24–27
2. Chambers R, Zweifach BW (1944) Topography and function of mesenteric capillary circulation. Am J Anat 75:173–205
3. Suzuki H, Schmid-Schönbein GW, Suematsu M, et al (1994) Impaired leukocyte-endothelial cell interaction in spontaneously hypertensive rats. Hypertension 24: 719–727
4. Suzuki H, Swei A, Zweifach BW, et al (1995) In vivo evidence for microvascular oxidative stress in spontaneously hypertensive rats: hydroethidine microfluorography. Hypertension 25:1083–1089
5. Suzuki H, Zweifach BW, Forrest MJ, et al (1995) Modification of leukocyte adhesion in spontaneously hypertensive rats by adrenal corticosteroids. J Leukocyte Biol 57:20–26
6. Suzuki H, Zweifach BW, Schmid-Schönbein GW (1995) Dependence of elevated mesenteric arteriolar tone on glucocorticoids in spontaneously hypertensive rats. Int J Microcirc 15:309–315
7. Suzuki H, Zweifach BW, Schmid-Schönbein GW (1995) Vasodilator response of mesenteric arterioles to histamine in spontaneously hypertensive rats. Hypertension 26: 397–400
8. Suzuki H, Zweifach BW, Schmid-Schönbein GW (1996) Glucocorticoid modulates vasodilator response of mesenteric arterioles in spontaneously hypertensive rats. Hypertension 27:114–118
9. Suzuki H, Poole DC, Zweifach BW, et al (1995) Temporal correlation between maximum tetanic force and cell death in postischemic rat skeletal muscle. J Clin Invest 96:2892–2897
10. Suzuki H, DeLano FA, Parks DA, et al (1998) Xanthine oxidase activity associated with arterial blood pressure in spontaneously hypertensive rats. Proc Natl Acad Sci USA 95:4754–4759
11. Suzuki H, Suematsu M, Schmid-Schönbein GW (1999) Microvascular oxidative stress, immune reaction and apoptosis in hypertensives. Clin Hemorheol Microcirc 21(3–4): 161–168
12. Suematsu M, Suzuki H, Delano FA, et al (2002) The inflammatory aspect of the microcirculation in hypertension: oxidative stress, leukocytes/endothelial interaction, apoptosis. Microcirculation 9(4):259–276
13. Warren JR, Marshall BJ (1983) Unidentified curved bacilli on gastric epithelium in active chronic gastritis. Lancet i:1273–1275
14. Suzuki H, Mori M, Seto K, et al (2001) Polaprezinc attenuates the *H. pylori*-induced gastric mucosal leukocyte activation in Mongolian gerbils. A study using intravital videomicroscopy. Aliment Pharmacol Ther 15:715–725
15. Suzuki H, Yamabe K, Ishii H (2000) *Helicobacter pylori:* current topics. Asian Med J 43:343–348
16. Suzuki H, Masaoka T, Miyazawa M, et al (2002) Gastric mucosal response to *Helicobacter pylori.* Keio J Med 51(suppl 2):40–44
17. Kalia N, Jones C, Bardhan DK, et al (2001) Effects of genotypically different strains of Helicobacter pylori on human microvascular endothelial cells in vitro. Dig Dis Sci 46(1):54–61

18. Jenkinson L, Bardhan KD, Atherton J, et al (2002) *Helicobacter pylori* prevents proliferative stage of angiogenesis in vitro: role of cytokines. Dig Dis Sci 47(8):1857–1862
19. Elizalde JI, Gomez J, Panes J (1997) Platelet activation in mice and human Helicobacter pylori infection. J Clin Invest 100:996–1005
20. Suzuki H, Mori M, Sakaguchi AA, et al (1998) Enhanced levels of C-X-C chemokine, human GROα, in *H. pylori*-associated gastric disease. J Gastroenterol Hepatol 13:516–520
21. Suzuki M, Miura S, Suematsu M, et al (1992) *Helicobacter pylori*-associated ammonia production enhances neutrophil-dependent gastric mucosal cell injury. Am J Physiol 263:G719–G725
22. Suzuki H, Mori M, Suzuki M, et al (1997) Extensive DNA damage induced by monochloramine in gastric cells. Cancer Lett 115:243–248
23. Suzuki H, Seto K, Mori M, et al (1998) Monochloramine (NH_2Cl) induced DNA fragmentation in gastric cell line, MKN 45. Am J Physiol 275:G712–G716
24. Kiss J, Lamarque D, Moran AP, et al (2001) *Helicobacter pylori* lipopolysaccharide-provoked injury to rat gastroduodenal microvasculature involves inducible nitric oxide synthase. Eur J Pharmacol 420(2–3):175–179
25. Whittle BJ, Morschl E, Pozsar J, et al (2001) *Helicobacter pylori* lipopolysac charide provokes iNOS-mediated acute systemic microvascular inflammatory responses in rat cardiac, hepatic, renal and pulmonary tissues. J Physiol (Paris) 95(1–6):257–259
26. Suzuki H, Mori M, Seto K, et al (1999) *H. pylori*-associated gastric pro- and anti-oxidant formation in Mongolian gerbils. Free Radic Biol Med 26:679–684
27. Yoshida M, Wakabayashi G, Ishikawa H, et al (2002) Arteriovenous shunting blood flow is intravitally observed in the stomach after thermal injury in rats. Keio J Med 51(4):193–200
28. Kojima M, Hosoda H, Date Y, et al (1999) Ghrelin is a growth-hormone-releasing acylated peptide from stomach. Nature 402:656–660
29. Dornonville de la Cour C, Bjorkqvist M, Sandvik AK, et al (2001) A-like cells in the rat stomach contain ghrelin and do not operate under gastrin control. Regul Pept 99:141–150
30. Nakazato M, Murakami N, Date Y, et al (2001) A role for ghrelin in the central regulation of feeding. Nature 409:194–198
31. Suzuki H, Masaoka T, Hosoda H, et al (2004) *Helicobacter pylori* infection modifies gastric and plasma ghrelin dynamics in Mongolian gerbils. Gut 53(2):187–194
32. Nwokolo CU, Freshwater DA, O'Hare P, et al (2003) Plasma ghrelin following cure of *Helicobacter pylori*. Gut 52(5):637–640
33. de Luis DA, Lahera M, Canton R, et al (1998) Association of *Helicobacter pylori* infection with cardiovascular and cerebrovascular disease in diabetic patients. Diabetes Care 21(7):1129–1132
34. Quadri R, Rossi C, Catalfamo E, et al (2000) *Helicobacter pylori* infection in type 2 diabetic patients. Nutr Metab Cardiovasc Dis 10(5):263–266
35. Warzecha Z, Dembinski A, Ceranowicz P, et al (2002) Deleterious effect of Helicobacter pylori infection on the course of acute pancreatitis in rats. Pancreatology 2(4):386–395

Ischemia/Reperfusion Injury in the Stomach: Role of Oxygen-Derived Free Radicals and Complement Regulatory System

Takashi Joh, Tadayuki Oshima, Satoshi Tanida, Makoto Sasaki, Hiromi Kataoka, and Makoto Itoh

Key words. Ischemia/Reperfusion, Free radical, Complement, ^{51}Cr-EDTA clearance, Decay-accelerating factor (DAF)

It is well recognized that ischemia/reperfusion (I/R) causes gastric epithelial damage during the reperfusion period by oxygen free radicals-mediated mechanism. This issue concerning stomach was first reported in the rat hemorrhagic shock model by Itoh and Guth in 1985 [1]. After this report, many scientists developed new experimental models to investigate further mechanisms for gastric I/R damage. We established a new local gastric I/R model [2], in which gastric epithelial damage was assessed by monitoring the blood-to-lumen clearance of ^{51}Cr-labeled ethylenediamine tetraacetic acid (^{51}Cr-EDTA). We also developed a guinea pig model to study healing process from gastric mucosal erosion caused by I/R [3]. Using these models, we have investigated many biological factors which may act on path-physiology of gastric I/R injury such mucous, NO, complements, and complement regulatory protein. In this chapter, we mention a little bit about history, and discuss the role of complement and the complement regulatory system in gastric mucosal damage caused by I/R.

Role of Oxygen-Derived Free Radicals in Gastric Ischemia/Reperfusion Injury

Figure 1 shows mucosal lesions which were induced in the first model for gastric I/R injury reported by Itoh and Guth in 1985 [1]. In this study, 0.1 N HCl (1 ml) was instilled into the pylorus-ligated stomach of the anesthetized

---

Department of Internal Medicine and Bioregulation, Nagoya City University Graduate School of Medical Sciences, 1 Kawasumi, Mizuho, Nagoya 467-8601, Japan

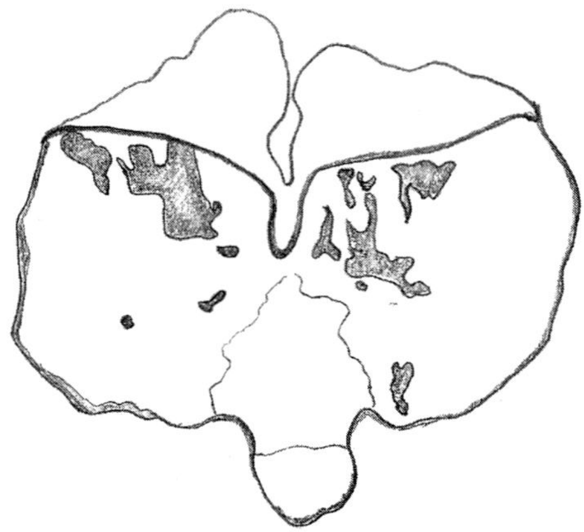

FIG. 1. Hemorrhagic shock-induced gastric lesions in the rat

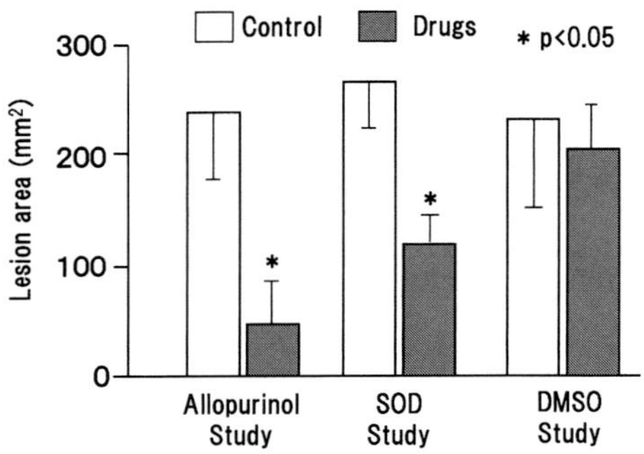

FIG. 2. Effects of oxygen radical scavengers on gastric ischemia/reperfusion (I/R) injury. *SOD*, superoxide dismutase; *DMSO*, dimethyl sulfoxide

rat, and then the rat was bled to reduce the blood pressure to less than 30 mmHg. The blood pressure was maintained at less than 30 mmHg for 20 min, and then the shed blood was retransfused. The area of gastric mucosal lesions was measured 20 min after the retransfusion (Fig. 1).

This study was designed to determine whether oxygen-derived free radicals play a role or not. For this purpose, allopurinol (an inhibitor of xanthine oxidase), superoxide dismutase (SOD; a scavenger of superoxide radicals), and dimethyl sulfoxide (DMSO; a scavenger of hydroxyl radicals) were used. Both allopurinol and SOD significantly protected against hemorrhagic shock-induced gastric lesions, although DMSO did not (Fig. 2). These results indicated that oxygen-derived free radicals produced in the xanthine/xanthine oxidase system [4] play an important role in the formation of gastric lesions produced by I/R plus HCl.

In 1994, we evaluated the damage to the gastric epithelium produced by local I/R without luminal instillation of 0.1 N HCl [2]. Local gastric ischemia was induced by clamping the left gastric artery which induced a 67% reduction in blood flow only in the corpus. No measurable gross lesion was observed in this model, but epithelial damage was clearly detected by measurement of the blood-to-lumen clearance of ^{51}Cr-EDTA. In the absence of

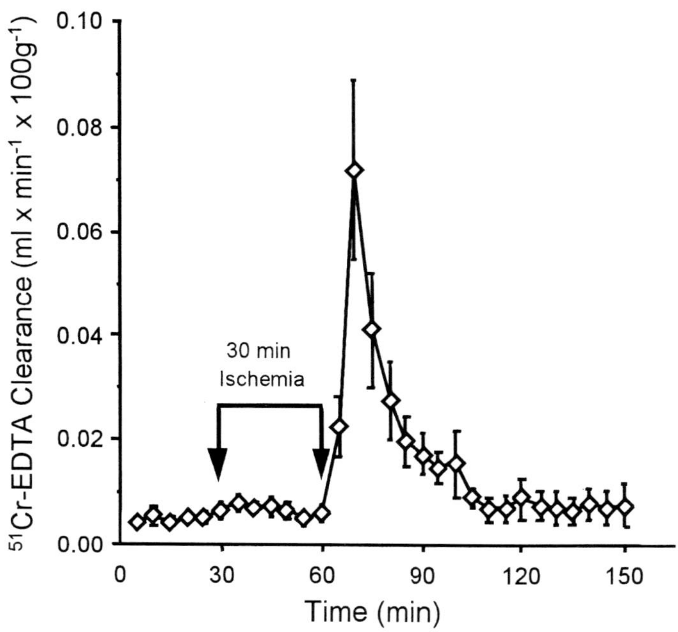

FIG. 3. Gastric epithelial damage induced by clamping left gastric artery

exogenous acid, the histological damage was also minimum and could not be quantified. However, a significant increase in ^{51}Cr-EDTA clearance was observed shortly after reperfusion in a manner that depended on the duration of ischemia. This increase in clearance reached a maximum approximately 10 min after reperfusion and returned rapidly toward control levels within 40–50 min after reperfusion in a manner consistent with the concept of restitution (Fig. 3). Allopurinol, SOD, and DMSO significantly reduced this leakage of ^{51}Cr-EDTA after gastric I/R. These results indicated that this model also supported the xanthine/xanthine oxidase hypothesis [4].

Role of Complement Regulatory System

It is known that I/R-related factors such as acidosis, platelet activating factor, and anoxia can activate the alternative pathway of complement, and that once activated, complement can cause tissue damage [5,6]. During the activation process, anaphylatoxins (C3a and C5a) are generated. These fragments of complement strongly activate neutrophils, bosophils`and mast cells.

An endogenous membrane-binding complement regulatory protein, decay-accelerating factor (DAF), protects host tissues from damage mediated by complement activation. Mucosal DAF is known to be upregulated in various inflammations including ulcerative colitis. However, gastric mucosal expression of DAF after ischemia has not been well investigated. We therefore evaluated the role of complement/complement regulatory factors in damage to the gastric mucosa caused by I/R [3–6].

In these experiments, male Hartley guinea pigs (250–300 g) were used. Animals were fasted overnight and anesthetized. Gastric ischemia was induced by clamping the left gastric and epiploic arteries for 30 min. Prior to sacrifice, stomachs were removed 0, 6, 12, 24 h, and 3 and 7 days after I/R. Complement was suppressed by intraperitoneal administration of cobra venom factor (CVF). Pretreatment with CVF (50 units) prior to I/R reduced the serum complement value (CH_{50}) to an undetectable level within 20–24 h of injection. Immunohistochemical studies of DAF, were performed with a specific monoclonal antibody to guinea pig DAF after local gastric I/R. In guinea pigs, alternative splicing of DAF mRNA generates both glycosylphosphatidylinositol (GPI)-anchored types and transmembrane types (TM) of DAF (Fig. 4). The DAF mRNA and its isoforms in gastric tissues after I/R were examined by reverse transcription–polymerase chain reaction and Northern blotting analysis. Localization of gastric DAF mRNA expression after I/R was also evaluated by in situ hybridization.

To investigate isoforms expressed in the normal gastric epithelium by DAF, we separated epithelium from the remnant stromal tissue using a crypt isolation procedure [7]. Reverse transcription–polymerase chain reaction demonstrated that the GPI-anchored isoforms are predominant over the TM

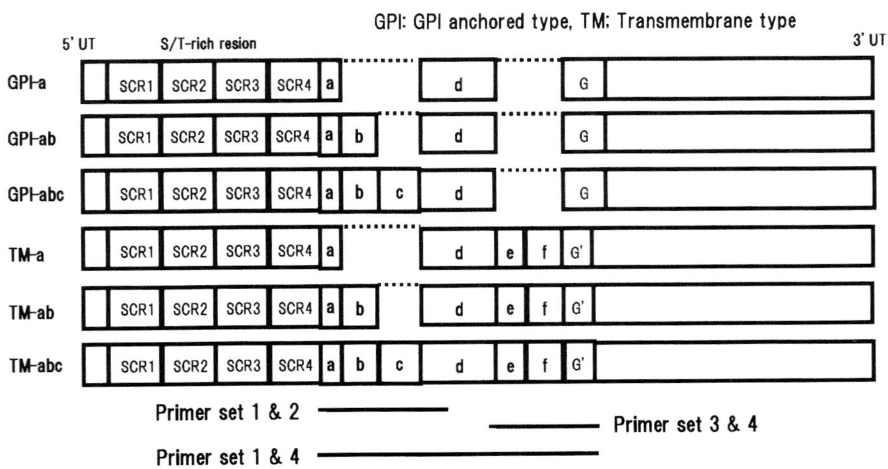

FIG. 4. mRNAs for guinea pig decay-accelerating factor (*DAF*) isoforms and primer sets. *GPI*, glycosylphosphatidylinositol; *UT*, untranslated

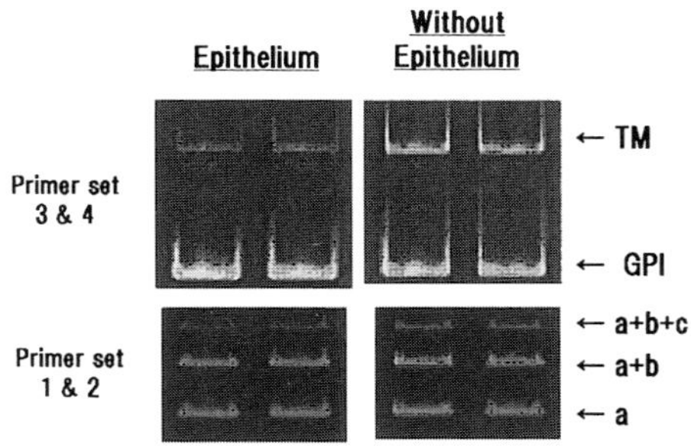

FIG. 5. Gastric DAF isoforms analyzed by reverse transcription–polymerase chain reaction. *TM*, transmembrane type; *GPI*, glycosylphosphatidylinositol anchored type

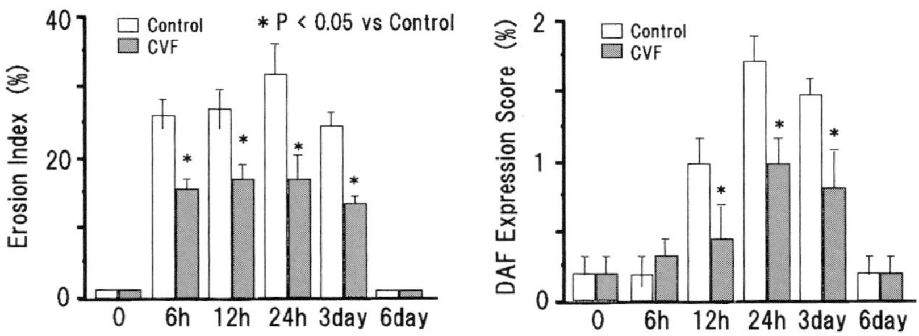

FIG. 6. Changes in gastric erosion and DAF expression after I/R. *CVF*, cobra venom factor

isoforms in all digestive epithelium. However, there was no difference between GPI-anchored type and TM type in the stromal tissue (Fig. 5).

Decay-accelerating factor expression was transiently upregulated after I/R. The DAF mRNA level in gastric tissues was found to be highest at 6h after I/R, returning to the baseline at 24h (Fig. 6). The DAF protein in gastric tissues was highest at 24h after I/R, returning to the baseline 6–7 days after I/R. Strong DAF mRNA expression detected by in situ hybridization was observed in the cytoplasm of cells beneath the eroded tissues 6h after I/R (Fig. 7). Reverse transcription–polymerase chain reaction analysis revealed that mRNAs of the TM types had become significantly dominant by 6h after I/R, while levels of the GPI-anchored types remained unchanged (Fig. 8). In guinea pigs depleted of complement by CVF treatment, the area of erosion and the upregulation of DAF expression after I/R were significantly limited (Fig. 6).

In summary, these data indicated that complement may play an important role in gastric I/R damage via their direct cytotoxic effects or their activation of white blood cells. The GPI-anchored form of DAF may play a protective role in normal gastrointestinal epithelial cells, while the TM-isoform of DAF might play an important role as an acute phase reactant at the site of inflammation in guinea pigs.

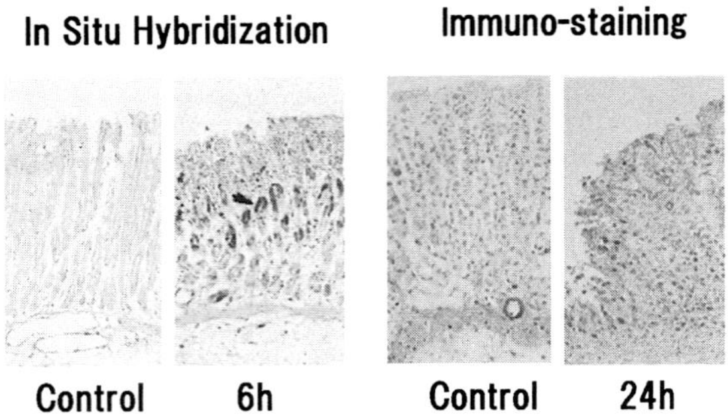

FIG. 7. Changes in DAF expression after I/R

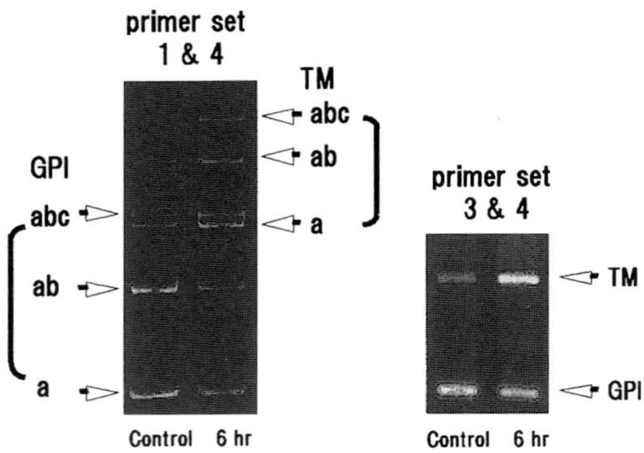

FIG. 8. Decay-accelerating factor isoform analysis after ischemia. *TM*, transmembrane type; *GPI*, glycosylphosphatidylinositol anchored type

References

1. Itoh M, Guth PH (1985) Role of oxygen-derived free radicals in hemorrhagic shock-induced gastric lesions in the rat. Gastroenterology 88:1162–1167
2. Kawai T, Joh T, Iwata F, et al (1994) Gastric epithelial damage induced by local I/R with or without exogenous acid. Am J Physiol 266:G263–G270
3. Oshima T, Okada N, Joh T, et al (2000) Decay-accelerating factor in guinea pig stomachs following ischemia reperfusion stress. J Immunol 164:1078–1085
4. Granger DN, Rutili G, McCord JM (1981) Superoxide radicals in feline intestinal ischemia. Gastroenterology 81:22–29
5. Ikai M, Itoh M, Joh T, et al (1996) Complement plays an essential role in shock following intestinal ischaemia in rats. Clin Exp Immunol 106:156–159
6. Joh T, Ikai M, Oshima T, et al (2001) Complement plays an important role in gastric mucosal damage induced by ischemia/reperfusion in rats. Life Sci 70:109–117
7. Mizoshita T, Joh T, Oshima T, et al (2002) Differential expression of decay-accelerating factor isoforms in the digestive tract of guinea pig. Life Sci 70:867–876

Gastroduodenal Microcirculatory Response to Luminal Acid

Jonathan D. Kaunitz[1,2,3], Shin Tanaka[4], and Yasutada Akiba[3]

Summary. The hyperemic response to luminal acid, a key protective mechanism for upper gastrointestinal mucosa, occurs by different mechanisms in the distal esophagus, stomach, and duodenum. The esophagus is a stratified squamous mucosa of high electrical resistance. Although luminal acid permeates only into the superficial epithelial layer, luminal acid induces protective mucosal hyperemia and clinical symptoms. The stomach has a well-studied microcirculatory response to luminal acid. Blood flow is believed to play a prominent role in mucosal protection, since the abolition of the hyperemic response to luminal acid is associated with enhanced mucosal injury susceptibility. Infusion of pentagastrin, mimicking the endogenous secretory gastric response to food, unmasked this hyperemic response in undamaged mucosa and activates neurons in the vagal nucleus. The duodenum is a leaky, low-resistance, columnar epithelium. Unlike the stomach, duodenal hyperemia is readily produced by perfusion with acidic luminal solutions. We further studied the afferent pathways involved with this hyperemic response. The vanilloid receptor (VR) antagonist capsazepine (CPZ) dose-dependently inhibited capsaicin-induced hyperemia. Capsazepine dose-dependently inhibited acid-induced hyperemia. The gastric and intestinal mucosae are richly innervated with VR. Each segment of the gastrointestinal tract exposed to gastric acid appears to have a unique protective mechanisms to which regulation of mucosal blood flow plays an important role.

---

[1]Greater Los Angeles Veteran Affairs Healthcare System, WLAVA Medical Center, [2]Department of Medicine, The David Geffen School of Medicine at UCLA, [3]CURE: Digestive Diseases Research Center, Building 114, Suite 217, 11301 Wilshire Boulevard, Los Angeles, CA 90073, USA

[4]Department of Internal Medicine, National Tokyo Medical Center, 2-5-1 Higashigaoka, Meguro-ku, Tokyo 152-8902, Japan

Key words. Stomach, Esophagus, Duodenum, Vanilloid receptor, Blood flow

Overview

The acid-exposed mucosa of the upper gastrointestinal tract is uniquely exposed to more than 7-log range of hydrogen ion concentrations. Gastric acid secretion, antral and fundic peristalsis, esophageal motility, and duodenal and pancreatic bicarbonate secretion all contribute to the varying pH profile to which the epithelial cells are exposed. The mucosal cells of the esophagus, stomach, and duodenum each are presented with unique challenges and solutions to the problem of defense against the ever-present acid stress to which they are exposed. In this chapter, we discuss how each organ copes with and defends against acid stress, emphasizing subepithelial microcirculatory mechanisms.

The microcirculation in the submucosa of the organs of the upper gastrointestinal tract plays an important role in the defense from injury to due luminal acid. The overall role of mucosal blood flow, apart from its customary and accepted role as supplier of oxygen and remover of CO_2 and cellular waste products, is somewhat controversial and thus far unproven. Certainly, the high energy requirements of gastric parietal cells dictate and constant circulation-delivered energy supply, but the function goes well beyond these considerations. The doctrine formulated by Silen [1] which suggested that the gastric microcirculation delivered bicarbonate and carried away excess acid equivalents continues to be the paradigm that serves as a useful framework for the interpretation of the data described in the following sections.

Organ-Specific Mechanisms

Esophagus

The esophagus is the least well studied organ in terms of defense mechanism from acid injury, even though acid-induced esophageal mucosal injury has become the most common upper gastrointestinal mucosal disease in North America [2]. The rabbit esophageal epithelium is electrically tight (1000–2500 $\Omega \cdot cm^2$), with contributory structural components located predominantly in the stratum corneum and upper stratum spinosum [3]. Unlike other acid-exposed organs, however, the replicating cells within the basal pre-epithelial layer are 25–30 cell layers removed from the luminal surface [3]. The esophageal mucosa is highly resistant to refluxed concentrated gastric acid [4,5]. Esophageal resistance to acid is thought to reside in luminal pH gradients that may result from bicarbonate secretion in some species, resistance to acid permeation due to high resistance intercellular structures, and intrinsic cellular resistance due to acid–base transporters in the basal cell plasma membrane [6,7]. Being a stratified squamous mucosa, the esophagus presumably resists acid injury due to its formidable multilayered structure and high intrinsic electrical resistance. Indeed, pre-epithelial factors such as a mucus coat and bicarbonate secretion are variably present or not, and intrinsic cellular acid-base transporters of the basal cell plasma membrane have been invoked [6,7]. Esophageal blood flow has been measured infrequently, with its contribution towards overall mucosal defense demonstrated in a few studies. For example, Bass and coworkers have shown that capsaicin-sensitive afferent nerves and calcitonin gene-related peptide (CGRP) mediate protective hyperemia [8,9]. We have found that luminal acid fails to permeate deeply into esophageal mucosa, with measured penetration only into the stratum corneum [10]. Coupled with the increase of blood flow in response to luminal acid perfusion (Fig. 1), these data are consistent with mucosal responses to luminal acid being mediated by neural acid-sensing mechanisms, as has been observed previously with stomach and duodenum, [11,12], or by non-neural pathways, rather than from direct penetration of acid through the epithelium

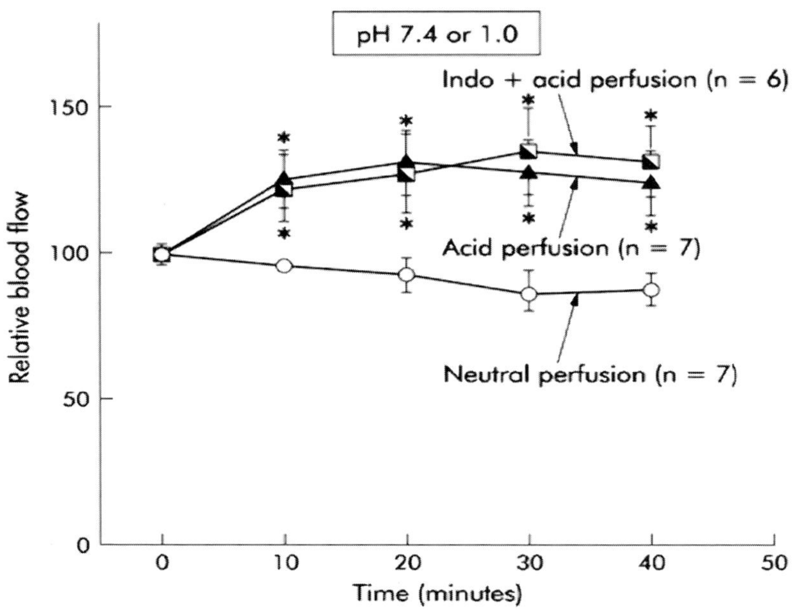

FIG. 1. Relative blood flow of esophageal mucosa. Acid perfusion significantly increased mucosal blood flow but indomethacin (*Indo*) did not affect this acid-induced increase. *$P < 0.05$ vs pH 7 perfusion by analysis of variance. Adapted from [10]

into the basal pre-epithelial layer. These sensing mechanisms have been suspected on the basis of the ability of subjects with reflux disease to sense the presence of acid perfused into the esophageal lumen [13]. Immunohistochemical data have not been consistent with penetration of afferent nerves into epithelial strata more superficial than the basal cell layer [14]. This raises the question as to how a luminal acid signal is transduced in the absence of proton permeation through the mucosa to the location of the afferent nerves. There are currently only hypotheses about how this can be accomplished, including direct cell-to-cell communication throughout the epithelial layer, or possible paracrine signaling mechanisms.

Stomach

The stomach is lined by a single layer of columnar epithelial cells, arranged into distinct glands populated by multiple cell types. Intrinsic resistance of the gastric mucosa is high, usually measured in the range of $1000\,\Omega\cdot cm^2$. The gastric microcirculation is the best studied of the gastrointestinal acid-exposed organs. Due to the proximity of acid secreting parietal cells, gastric luminal pH is higher than that of the adjacent duodenum or esopahagus. Hyperemia can be induced by luminal exposure to concentrated acid, but only in the presence of mucosal injury, that decreases the high intrinsic impermeability of the gastric epithelium [15]. In the stomach, for example, interventions that attenuate the hyperemic response to acid perfusion increase mucosal injury [16–18]. We have previously found that gastric hyperemia can also be induced in the presence of luminal acid in the presence of exogenous pentagastrin, simulating the postprandial gastric acid response, as shown in Fig. 2 [19]. Interruption of this hyperemic response enhances mucosal injury susceptibility [20], underscoring and confirming the important role played by the gastric microcirculation in mucosal defense. A recent paper from Holzer's laboratory has placed many of these findings in perspective. They found that c-fos was enhanced in the brains of rats whose stomachs were luminally exposed to acid in the presence of pentagastrin, confirming the role of vagal afferent pathways in the hyperemic response [21].

Duodenum

The proximal duodenum is exposed to cyclical and rapid variations of luminal pH. Unlike other acid-exposed organs such as the stomach or esophagus, the duodenum has high transepithelial permeability to water and solutes, necessitating the presence of nonstructural defense mechanisms such as mucus and bicarbonate secretion, and submucosal blood flow. We have shown previously that a brief exposure to intense luminal acidity, corresponding to physiological acid stress, enhances all measured duodenal defense mechanisms, includ-

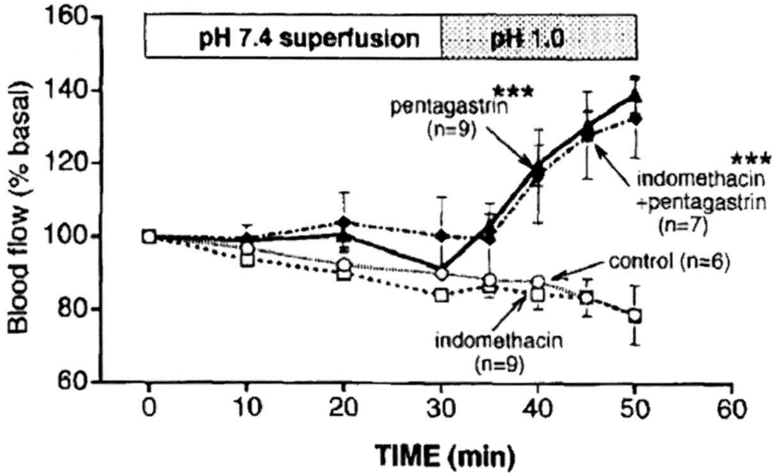

FIG. 2. Gastric blood flow in the presence of luminal acid, pentagastrin, and indomethacin. Blood flow was measured by laser-Doppler flowmetry. Mucosal blood flow in undamaged mucosa increased during perfusion of acid only in rats infused with pentagastrin. Adapted from [10]

ing mucus secretion/gel thickness increase, increased cellular bicarbonate concentration, and increased mucosal blood flow. The mucosal sensor underlying these rapid changes, however, remains unknown, although there are good data that indicate that the acid sensor is a component of the well-known afferent branch of the enteric nervous system, with actual acid sensing transduced by a newly discovered acid-sensitive receptor. Recent studies using vanilloid receptor (VR)-1 antibodies have revealed that there is intense staining of VR-1-immunoreactive nerves in the duodenal epithelium, including the lamina propria mucosa up to the villous tips and down to the pericryptal regions, the submucosal layer, and intrinsic nerves (myenteric plexus) [22]. These VR-1-positive nerves highly colocalize with CGRP. Furthermore, VR-1-positive neurons are present not only in the dorsal root ganglion (splanchnic afferent center), but also in the nodose ganglion (vagal afferent center) and in the myenteric plexus (intrinsic afferent center), and vagotomy, not sympathectomy abolishes the acid-induced hyperemic response in duodenum, suggesting that VR signaling projects to vagal afferents and intrinsic afferents, unlikely to splanchnic afferents [23]. These histological and surgical studies confirm the physiological observations reported above and have helped confirm our supposition regarding the nature of the acid-sensing protective upregulation of duodenal defense mechanisms, differently from the gastric defenses in which the splanchnic afferents contribute to the acid-induced hyperemia [24].

Our studies revealed some novel observations about the nature of duodenal blood flow and its regulation. For example, inhibition of sodium–proton exchange (NHE) with the potent amiloride analog dimethylamiloride inhibited the hyperemic acid response [25]. Interestingly, acidification of the cytoplasm by alternate means, such as with an ammonium chloride prepulse or valinomycin increased blood flow, also inhibitable by dimethylamiloride [25]. These studies suggested that acid must pass through the epithelial cell and exit via by the basolateral NHE isoform NHE1 prior to eliciting a hyperemic response. In further studies, we examined the sensing mechanisms underling the hyperemic response. Capsazepine, an antagonist to the recently cloned vanilloid receptor, abolished the hyperemic response to acid, confirming the involvement of vanilloid receptors in the acid response (Fig. 3). Further studies also confirmed that the hyperemic response was mediated by a well-known pathway that includes afferent sensory nerves and release of the neuropeptide CGRP and the vasodilatory gas nitric oxide, but was not inhibited by indomethacin, a nonselective inhibitor of cyclooxygenase [25]. These studies provided data supporting our proposed mechanism of duodenal acid-

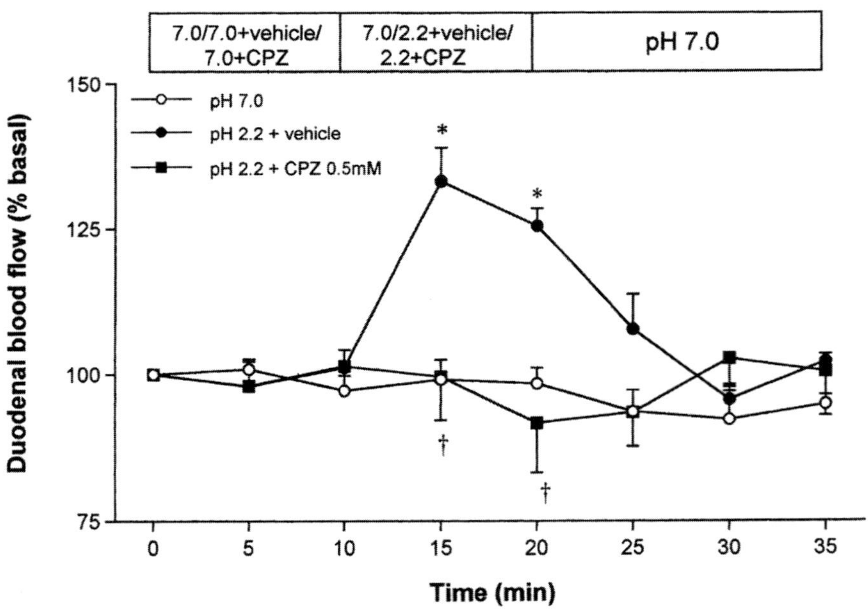

FIG. 3. Effect of CPZ on acid-induced hyperemia in duodenum. Acid perfusion (*closed circles*) rapidly increases duodenal blood flow compared with pH 7.0 alone (*open circles*) and returns to baseline after acid removal. Capsazepine (*CPZ*) (0.5 mM, *closed squares*) abolishes this hyperemic response to luminal acid. $^{*}P < 0.05$ vs pH 7.0 Krebs perfusion; $P < 0.05$ vs pH 2.2+vehicle perfusion. Data are means ± SE from six rats. Adapted from [12]

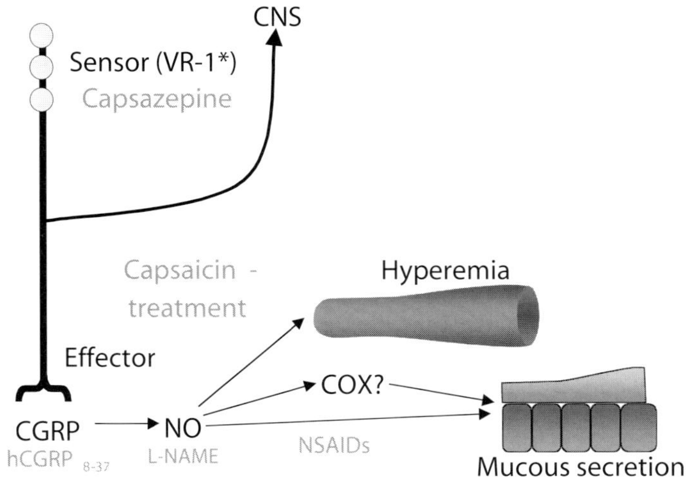

FIG. 4. The capsaicin pathway. We have depicted a scheme for the regulation of upper gastrointestinal defense mechanisms in response to luminal acid. Inhibitors are depicted in *gray*. *L-NAME*, N^G-nitro-*L*-arginine methyl ester; $hCGRP_{8-37}$, inhibitor of human calcitonin gene-related peptide; *NSAIDs*, nonsteroidal anti-inflammatory drugs; *COX*, cyclooxygenase; *VR*, vanilloid receptor; *CNS*, central nervous system; *NO*, nitric oxide

induced hyperemia, including acid diffusion into the epithelial cell, basolateral extrusion via NHE1, activation of vanilloid receptors on afferent nerves, CGRP release, with activation of endothelial nitric oxide synthesis, with production of vasodilatory nitric oxide. A scheme of proposed regulatory mechanisms for blood flow and mucous secretion is shown in Fig. 4.

Clinical Correlates

Upper gastrointestinal visceral acid sensors have received far less attention than have other chemosensors and distension sensors, despite the ubiquity of luminal acid and the importance for constant vigilance against acid-induced injury. Nevertheless, the presence of esophageal mucosal acid sensors has been inferred from the results of the Bernstein test, in which the esophageal mucosa is exposed to physiological concentrations of HCl in heartburn sufferers [13,26]. Even in the absence of mucosal damage, some can sense the presence of this acid, implying the presence of a transmucosal mechanism whereby luminal acid is transduced into an afferent neural signal. Little is known, however, about the contribution of these afferent signals, and the resulting hyperemia, towards prevention of mucosal injury. One admonition that was commonplace before the clinical development of potent antisecretory medicine was the abolition of spicy foods from the diet. Most spice derives its “heat” from red peppers, the active ingredient being capsaicin, a well-recognized ligand for VR-1. Perhaps the more recent finding that the ingestion of spicy food has no bearing on an underlying ulcer diathesis underscores this point; conversely by enhancing protective mechanism, capsaicin and thus spicy foods might actually be beneficial. One caveat is that repeated ingestion of heavily spiced foods could de-afferent the gastrointestinal tract, interrupting the capsaicin pathway, and perhaps increasing injury susceptibility. Finally, a duodenal acid load has been associated with nausea, bloating, and other visceral sensations identified with the irritable bowel syndrome [27]. Perhaps components of the capsaicin pathway are responsible for transducing luminal chemical stimuli into these neural signals.

Conclusions

Hyperemia is present in all gastrointestinal organs exposed to gastric acid. This response appears to be related to a combined sensing and efferent “capsaicin pathway” involving VR-1 and either vagal or intrinsic afferent nerves as the sensory component and CGRP/nitric oxide release as the efferent effectors. The capsaicin pathway appears to be responsible for the transduction of luminal chemical information (acid concentration) into neural signals, that produce symptoms such as heartburn, bloating, and nausea, and, more importantly, signal protective submucosal responses, in particular hyperemia.

Acknowledgments. Supported by VA Merit Review funding and NIH/NIDDK RO1 54221.

References

1. Dorricott NJ, Fiddian-Green RG, et al (1975) Mechanisms of acid disposal in canine duodenum. Am J Physiol 228:269–275
2. Heidelbaugh JJ, Nostrant TT, Kim C, et al (2003) Management of gastroesophageal reflux disease. Am Fam Physician 68:1311–1318
3. Orlando RC, Lacy ER, Tobey NA, et al (1992) Barriers to paracellular permeability in rabbit esophageal epithelium. Gastroenterology 102:910–923
4. Bateson MC, Hopwood D, Milne G, et al (1981) Oesophageal epithelial ultrastructure after incubation with gastrointestinal fluids and their components. J Pathol 133:33–51
5. Salo JA, Lehto VP, Kivilaakso E (1983) Morphological alterations in experimental esophagitis. Light microscopic and scanning and transmission electron microscopic study. Dig Dis Sci 28:440–448

6. Sarosiek J, McCallum RW (2000) Mechanisms of oesophageal mucosal defence. Baillieres Best Pract Res Clin Gastroenterol 14:701–717
7. Abdulnour-Nakhoul S, Nakhoul NL, Orlando RC (2000) Lumen-to-surface pH gradients in opossum and rabbit esophagi: role of submucosal glands. Am J Physiol Gastrointest Liver Physiol 278:G113–G120
8. Bass BL, Trad KS, Harmon JW, et al (1991) Capsaicin-sensitive nerves mediate esophageal mucosal protection. Surgery 110:419–425
9. McKie LD, Dunkin BJ, Pennanen MF, et al (1994) Esophageal mucosal blood flow: a central role for calcitonin gene-related peptide. Surgery 116:409–417
10. Tanaka S, Chu S, Hirokawa M, et al (2003) Direct measurement of acid permeation into rat oesophagus. Gut 52:775–783
11. Akiba Y, Furukawa O, Guth PH, et al (2001) Sensory pathways and cyclooxygenase regulate mucus gel thickness in rat duodenum. Am J Physiol Gastrointest Liver Physiol 280:G470–G474
12. Akiba Y, Guth PH, Engel E, et al (1999) Acid-sensing pathways of rat duodenum. Am J Physiol 277:G268–G274
13. Shi G, Tatum RP, Joehl RJ, et al (1999) Esophageal sensitivity and symptom perception in gastroesophageal reflux disease. Curr Gastroenterol Rep 1:214–219
14. Dutsch M, Eichhorn U, Worl J, et al (1998) Vagal and spinal afferent innervation of the rat esophagus: a combined retrograde tracing and immunocytochemical study with special emphasis on calcium-binding proteins. J Comp Neurol 398:289–307
15. Holzer P, Livingston EH, Guth PH (1991) Sensory neurons signal for an increase in rat gastric mucosal blood flow in the face of pending acid injury. Gastroenterology 101(2): 416–423
16. Holzer P, Livingston EH, Saria A, et al (1991) Sensory neurons mediate protective vasodilatation in rat gastric mucosa. Am J Physiol 60:G363–G370
17. Merchant NB, Dempsey DT, Grabowski MW, et al (1994) Capsaicin-induced gastric mucosal hyperemia and protection: the role of calcitonin gene-related peptide. Surgery 116:419–425
18. Li DS, Raybould HE, Quintero E, et al (1992) Calcitonin gene-related peptide mediates the gastric hyperemic response to acid back-diffusion. Gastroenterology 102:1124–1128
19. Tanaka S, Akiba Y, Kaunitz JD (1998) Pentagastrin gastroprotection against acid is related to H_2 receptor activation but not acid secretion. Gut 43:334–341
20. Holzer P, Pabst MA, Lippe IT, et al (1990) Afferent nerve-mediated protection against deep mucosal damage in the rat stomach. Gastroenterology 98:838–848
21. Danzer M, Jocic M, Samberger C, et al (2004) Stomach-brain communication by vagal afferents in response to luminal acid backdiffusion, gastrin and gastric acid secretion. Am J Physiol Gastrointest Liver Physiol 286(3):G403–G411
22. Akiba Y, Nakamura M, Nagata H, et al (2002) Acid-sensing pathways in rat gastrointestinal mucosa. J Gastroenterol 37(suppl) 14:133–138
23. Akiba Y, Nakamura M, Ishii H (2003) Acid sensing pathways of hypermic repsone to luminal acid involves vagal afferents and vanilloid receptor in rat dudoenum (abstract). Gastroenterology 124:A613–A614
24. Holzer P (2001) Gastroduodenal mucosal defense: coordination by a network of messengers and mediators. Curr Opin Gastroenterol 17:489–496
25. Akiba Y, Kaunitz JD (1999) Regulation of intracellular pH and blood flow in rat duodenal epithelium in vivo. Am J Physiol 276:G293–G302

26. Fass R, Yalam JM, Camargo L, et al (1997) Increased esophageal chemoreceptor sensitivity to acid in patients after successful reversal of Barrett's esophagus. Dig Dis Sci 42:1853–1858
27. Samsom M, Verhagen MA, vanBerge Henegouwen GP, et al (1999) Abnormal clearance of exogenous acid and increased acid sensitivity of the proximal duodenum in dyspeptic patients. Gastroenterology 116:515–520

Intracellular pH (pH_i) Measurement in Rat Duodenal Epithelium In Vivo Using SNARF-1

Tetsu Takeuchi, Lening Zhang, and Jonathan D. Kaunitz

Key words. SNARF-1, pH_i, Duodenal epithelium, In vivo, Acid

Introduction

Our laboratory has previously used BCECF to measure the pH_i of the duodenal epithelium [1–3]. BCECF loads into epithelial cells and fluoresces well, but since BCECF is a dual excitation-single emission compound, a ratio pair cannot be obtained simultaneously, but requires sequential acquisition of image pairs. Measurement of pH_i in vivo is impaired, due to the constant motion of epithelial cells which are affected by respiratory and cardiac movement. Exact registration of image pairs is not possible and ratiometric pH_i calculations must be therefore be performed manually. 5-(and 6)-Carboxy-seminaphthorhodafluor-1 (SNARF-1) is a single excitation dual emission fluorescent compound. The emission ratios calculated from the two peaks of 640 and 580 nm are a sensitive indicator of pH; furthermore, image pairs at two different wavelengths can be obtained simultaneously. It is thought that this specific feature can be used for more accurate determination of pH_i especially in vivo. In the present paper, we used SNARF-1 for measurement of pH_i in rat duodenal epithelium in vivo.

Material and Methods

All studies were approved by the Animal Use Committee of the West Los Angeles Veterans Affairs Medical Center. An in vivo microfluorometric technique was adapted from a technique originally developed by Tanaka et al. [1]

West Los Angeles VA Medical Center, Building 114, Suite 217, CURE: Digestive Diseases Research Center, and UCLA School of Medicine, Los Angeles, CA 90073, USA to measure pH_i in rat duodenal epithelial cells. Fluorescence of the duodenal mucosa observed microscopically was excited at 488 nm and divided into image pairs using the MicroImager with two emission filters (580/640 nm). The entire duration of the experiment was 40 min and each image pair was captured every 2 min for 15 min from the beginning of acid exposure (every 5 min for other observations). Three small areas of the duodenal epithelium from each recorded image pair were selected at random and each intensity of fluorescence was measured using image analyzer software (OpenLab, Improvision, Lexington, MA, USA).

Results

An examination of mucosa loaded with SNARF-1 in vivo produced image pairs such as depicted in Fig. 1. Duration of fluorescence and brightness was about 60 min. Note that fluorescence appears to be limited to the villous epithelial cells. Similar image pairs were obtained simultaneously at different wavelengths (580/640 nm). After baseline stabilization, pH 2.5 perfusion for 5 min rapidly reduced the pH_i. After acid removal, pH_i increased to baseline.

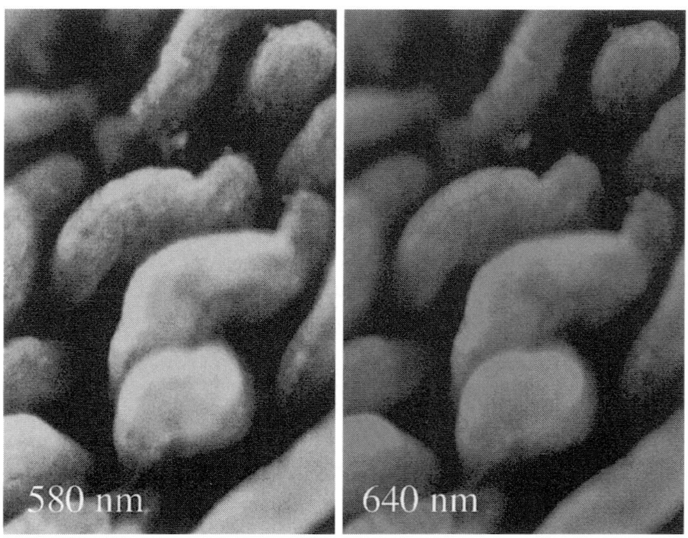

Fig. 1. In vivo image of chloromethyl SNARF-1-loaded duodenal mucosa. Identical image pairs are obtained simultaneously at two different wavelengths (580/640 nm). The dye localizes only in epithelial cells of villous tips

Discussion

pH_i measurement in vitro with SNARF-1 has been recently published [4,5]. Simultaneously acquired image pairs obtained with SNARF-1 enabled us to measure pH_i in vivo more precisely. Image pairs obtained every 2 min showed more accurate changes of pH_i with acid exposure than measurements made using BCECF. Since our experiment needed at least 40 min to complete, we used chloromethyl SNARF-1 acetate. It is designed for long-term cellular retention via coupling to protein thiols. In summary, we demonstrated that SNARF-1 provides a useful alternative means for the measurement of duodenal epithelial pH_i in vivo and hereafter, as the disadvantages of SNARF-1 are overcome, this dye can be applied to other organs or species.

Acknowledgment. Supported by VA Merit Review funding and NIH/NIDDK RO1 54221.

References

1. Tanaka S, Guth PH, Engel E, et al (1998) A novel technique for the in vivo measurement of intracellular pH (pH_i) of rat duodenal epithelial cells (abstract). Gastroenterology 114:A306
2. Akiba Y, Kaunitz JD (1999) Regulation of intracellular pH and blood flow in rat duodenal epithelium in vivo. Am J Physiol. 276(1 Pt 1):G293–G302
3. Akiba Y, Furukawa O, Guth PH, et al (2001) Cellular bicarbonate protects rat duodenal mucosa from acid-induced injury. J Clin Invest 108(12):1807–1816
4. Bassnett S, Reinisch L, Beebe DC (1990) Intracellular pH measurement using single excitation-dual emission fluorescence ratios. Am J Physiol 258(1 Pt 1):C171–C178
5. Blank PS, Silverman HS, Chung OY, et al (1992) Cytosolic pH measurements in single cardiac myocytes using carboxy-seminaphthorhodafluor-1. Am J Physiol. 263(1 Pt 2): H276–H284

Prevention of Ethanol-induced Gastric Mucosal Microcirculatory Disturbances by Mild Irritant Through the Actions of Calcitonin Gene-Related Peptide and Prostaglandin I_2 in Rats

Takeo Saeki[1], Takashi Ohno[2], Kazuhisa Kamata[1], Katsuharu Arai[1], Sumito Mizuguchi[1], Kouichiro Hattori[1], Katsunori Saigenji[1], and Masataka Majima[3]

Key words. Microcirculation, Gastric mucosal blood flow, Calcitonin gene-related peptide, Prostaglandin I_2, Sodium chloride, Ethanol

Introduction

Disturbances in the gastric mucosal microcirculation are thought to be an important cause of injury, and observation of the microcirculation is therefore important for revealing the pathophysiology of that injury. In the present study, we investigated the mechanism of adaptive cytoprotection of 1 M NaCl against ethanol [1] by observing the microcirculation in the basal part of the gastric mucosa of rats using intravital microscopy [2–4] to clarify the mediators involved in the maintenance of the integrity of the mucosal blood flow.

Materials and Methods

The mucosal microcirculation of male Sprague-Dawley rats anesthetized with urethane was observed using the methods that we previously developed [2–4]. The microcirculation of the basal part of the mucosa was examined with a light microscope with a long working distance objective lens. One arteriole, one venule, and two collecting venules were selected in one observation

[1] Department of Internal Medicine, Kitasato University School of Medicine, Sagamihara, Kanagawa 228-8555, Japan
[2] Department of Internal Medicine, Isehara Kyodo Hospital, Isehara, Kanagawa 259-1132, Japan
[3] Departments of Pharmacology and Molecular Pharmacology, Kitasato University School of Medicine, Kitasato University Graduate School of Medical Sciences, Sagamihara, Kanagawa 228-8555, Japan

window, and their internal diameters were measured using an adjustable electronic microscaler. The diameters of the arterioles were expressed as percentages of the original diameter. Ethanol (50%; 1 ml) and NaCl (1 M; 1 ml) were placed between the chamber and the gastric mucosa. Calcitonin gene-related peptide (CGRP)-(8–37) (10 μM; 20 μl), prostaglandin E_2 (PGE_2; 1 nM–10 μM; 20 μl) and a prostaglandin I_2 (PGI_2) analogue, beraprost sodium (1 nM–10 μM; 20 μl) were applied in the window. Indomethacin (1 mg/kg; Banyu Pharmaceutical, Tokyo, Japan) was administered intravenously.

Results and Discussion

Microcirculatory Changes in the Basal Part of the Gastric Mucosa After Exposure of the Mucosa to 50% Ethanol

The gastric mucosa was exposed for 3 min to 50% ethanol, which was then washed out with Tyrode's solution. Application of 50% ethanol on the mucosal side caused marked dilatation of the arterioles, but the collecting venules and the venules were strongly constricted, the internal cross-section of the collecting venules in particular becoming very small (Fig. 1B).

When 1 M NaCl was applied to the mucosa 5 min before mucosal treatment with 50% ethanol, the ethanol-induced rapid constrictions of the collecting venules and venules were completely inhibited (Fig. 1D), although 1 M NaCl alone caused some constriction of both the collecting venules and the venules (Fig. 2). The diameter of the arterioles during ethanol exposure was not affected by prior administration of 1 M NaCl, although they were markedly dilated during 1 M NaCl treatment (Fig. 2).

Furthermore, to investigate whether or not this preventive effect exerted by 1 M NaCl was attributable to the release of endogenous PGs, indomethacin (10 mg/kg, i.v.) was infused before the application of 1 M NaCl. The ethanol-induced constriction of the collecting venules, which had been blocked by 1 M NaCl pretreatment, appeared again in rats treated with indomethacin. The dilatation of arterioles observed during ethanol exposure was not affected by indomethacin preapplication.

Next, we applied CGRP-(8–37) (10 μM), a CGRP receptor antagonist, to the microvasculature 3 min before treatment of 1 M NaCl to investigate the involvement of endogenous CGRP. The treatment with CGRP-(8–37) did not affect the size of microvessels under 1 M NaCl application, but the ethanol-induced constriction of the collecting venules, which had been blocked by 1 M NaCl pretreatment, appeared again on treatment with a CGRP antagonist. The dilatation of arterioles observed during ethanol exposure was not affected by the applications of CGRP-(8–37).

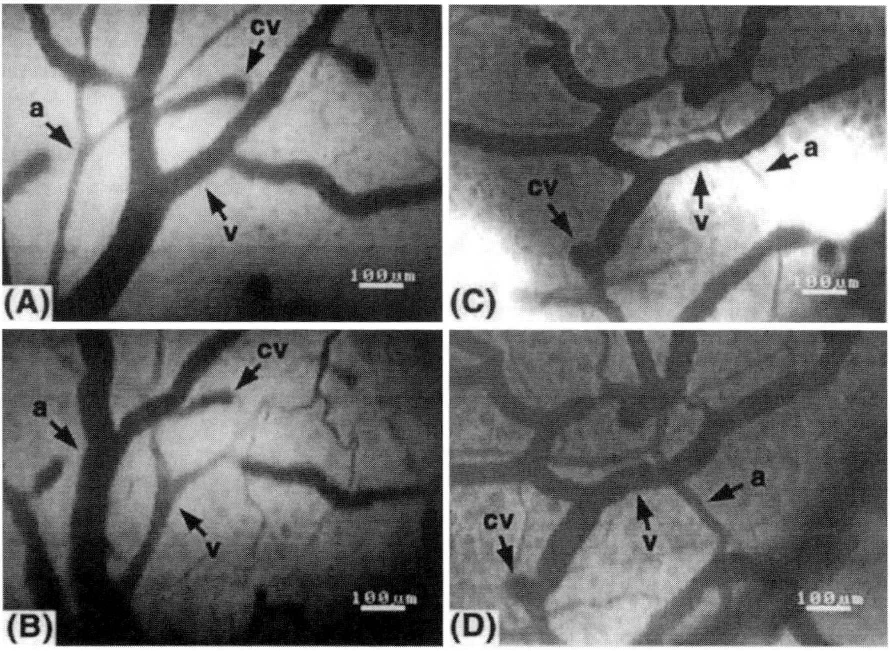

FIG. 1. Changes of the diameters of microvessels in the basal part of the mucosal microcirculation before (**A**) and 4 min after (**B**) application of 50% ethanol on the mucosal side of the stomach wall for 3 min. Marked dilatation of arterioles (*a*), and severe constriction of the collecting venules (*cv*) and the venules (*v*) in the gastric mucosa were observed under intravital microscopy. The diameters of microvessels in the basal part of the mucosal microcirculation before application of 1 M NaCl and ethanol (**C**). When 1 M NaCl was applied to the window 5 min before mucosal treatment with 50% ethanol, the ethanol-induced rapid constrictions of the collecting venules and venules were completely inhibited (**D**)

These results were similar to those obtained from indomethacin-pretreated rats. Thus, it is suggested that the protective effect of 1 M NaCl against ethanol-induced mucosal injury was mediated by both endogenous PGs and CGRP.

Effects of Administration of PGs (PGE_2 or Beraprost) on Ethanol-Induced Changes in the Microcirculation of the Basal Part of the Gastric Mucosa

Administration of PGE_2 (1 nM–10 μM) or beraprost (1 nM–10 μM) to the observation window caused a rapid and dose-dependent dilatation of the arterioles. PGE_2 caused a dose-dependent constriction of the collecting venules, but beraprost induced no changes in the diameter of the collecting venules. To identify the PGs contributing to 1 M NaCl-induced protective

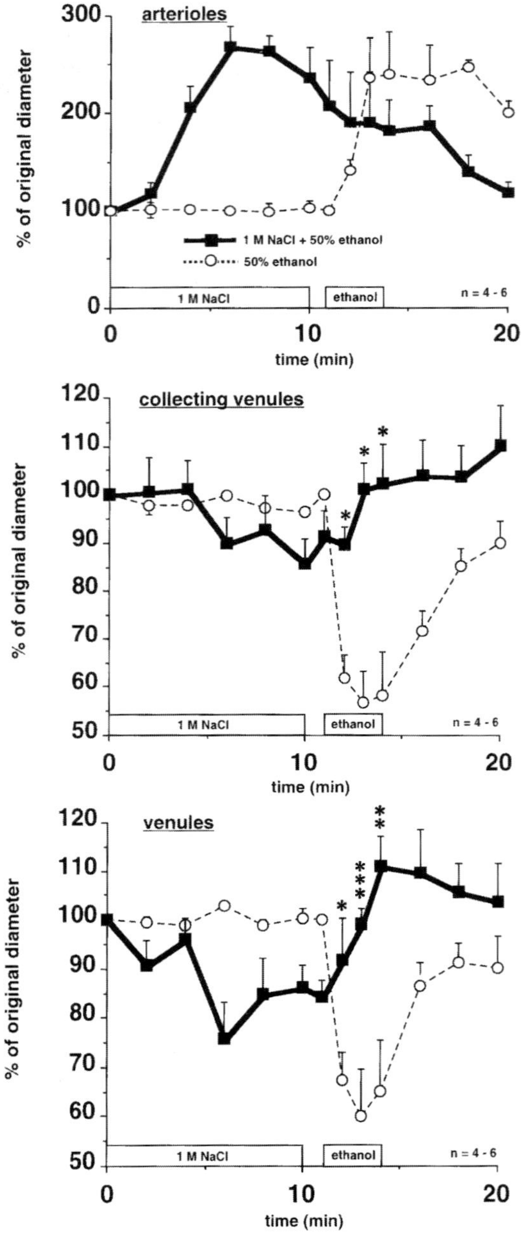

FIG. 2. Effects of prior application of 1 M NaCl on the ethanol-induced changes in the diameter of mucosal microvessels. The constrictions of the collecting venules (*middle panel*) and venules (*lower panel*) induced by 50% ethanol were markedly inhibited by 1 M NaCl pretreatment. The *upper panel* depicts results in arterioles. Each value indicates mean + SEM. *$p < 0.05$, **$p < 0.01$, ***$p < 0.001$

action, PGE_2 or beraprost was applied to the window 3 min before the mucosal application of 50% ethanol. The low doses of PGE_2 or beraprost which did not influence the original size of the gastric mucosal vessels including arterioles are given. Both PGE_2 (100 nM) and beraprost (1 nM) inhibited ethanol-induced rapid constrictions of the collecting venules. The dilatation of arterioles observed during ethanol exposure was not affected by applications of these doses of PGE_2 and beraprost.

To test whether or not the preventive effects of PGE_2 and beraprost were attributable to the increased release of endogenous CGRP, a sufficient dose of CGRP-(8–37) (10 μM) was applied to the window 3 min before the application of PGs (PGE_2 100 nM or beraprost 1 nM). The blockade of ethanol-induced constriction of the collecting venules by beraprost appeared again with the use of CGRP-(8–37). By contrast, that by PGE_2 did not appear even with CGRP-(8–37). These suggested that beraprost protected the gastric mucosa from ethanol through the cancellation of the constriction of collecting venules utilizing CGRP, but PGE_2 did offer protection independently, without CGRP.

In the present study, we found first that the mechanism of prevention by 1 M NaCl of ethanol-induced gastric mucosal injury is the inhibition of the constriction of collecting venules and venules, via endogenous PGs and CGRP. Second, it was seen that the PGI_2 analogue, which did not dilate the arterioles, collecting venules, or venules themselves, inhibited the ethanol-induced constriction of the collecting venules and venules, suggesting that endogenous PGI_2 was responsible for the preventive effect seen when 1 M NaCl was applied before the ethanol-induced injury through the increased release of CGRP. Thus, the dilator of the collecting venules, the venules or both may in future become a useful agent for preventing the induction of gastric mucosal injury by various necrotizing agents such as ethanol.

References

1. Boku K, Ohno T, Saeki T, et al (2001) Adaptive cytoprotection mediated by prostaglandin I_2 is attributable to sensitization of CGRP-containing sensory nerves. Gastroenterology 120:134–143
2. Ohno T, Katori M, Nishiyama K, et al (1995) Direct observation of microcirculation of the basal region of rat gastric mucosa. J Gastroenterol 30:557–564
3. Ohno T, Katori M, Majima M, et al (1999) Dilatation and constriction of rat gastric mucosal microvessels through prostaglandin EP2 and EP3 receptors. Aliment Pharmacol Ther 13:1243–1250
4. Saeki T, Ohno T, Boku K, et al (2000) Mechanism of prevention by capsaicin of ethanol-induced gastric mucosal injury: a study using intravital microscopy. Aliment Pharmacol Ther 14:135–144

Role of Perforin and Granzyme B of Cytotoxic T Lymphocyte in the Onset of Peptic Ulcer Formation

Tadashi Ohara[1], Tetsuo Morishita[1], Hidekazu Suzuki[2], Tatsuhiro Masaoka[1], and Hiromasa Ishii[2]

Key words. Cytotoxic T lymphocyte, Perforin, Granzyme B, *H. pylori*, Apoptosis, Peptic ulcer

Introduction

Attention has been focused on apoptosis in terms of its involvement in the pathogenesis of various diseases including hepatitis, graft-versus-host disease (GVHD), and multiple sclerosis [1–3]. Some pathways are known to be implicated in cytotoxic T lymphocyte (CTL)-mediated apoptotic induction: perforin/granzyme pathway (based on granules stored in cytoplasmic azurophile granules of CTLs), Fas/Fas ligand (FasL), tumor necrosis factor (TNF)/TNF receptor (TNFR), or TNF-related apoptosis-inducing ligand (TRAIL) pathways based on type II transmembrane proteins belonging to the TNF family expressed on the cell surface. Among them, the perforin/granzyme and Fas/FasL pathways are considered the two major routes of apoptosis. The aim of this study is to investigate whether apoptosis associated with peptic ulcer formation may occur via either or both of these two pathways.

Methodology

Patients

Thirty-two patients with active peptic ulcer at the A_1–A_2 stage with and without *Helicobacter pylori* infection were enrolled this study (gastric ulcer: 21; duodenal ulcer: 11). *Helicobacter pylori* infection was diagnosed by enzyme-linked immunosorbent assay and the urea breath test (UBT) method.

[1] Department of Internal Medicine, Tokyo Dental College, Chiba, Japan
[2] Department of Internal Medicine, Keio University School of Medicine, 35 Shinanomachi, Shinjuku-ku, Tokyo 160-8582, Japan

Methods

Materials were biopsy specimens which taken from patients endoscopically. Experimental protocol was as follows: (1) detection of apoptotic cells by in situ fluorescence (TUNEL method), (2) examination of the induction of perforin, Fas, and Fas ligand (FasL) by reverse transcription–polymerase chain reaction, (3) examination of the induction of caspase 3 by immunoblotting, and (4) examination of the detection of perforin, granzyme B, and caspase 3 by immunohistochemistry (light and electron microscopy staining).

Results

Expression of apoptotic cells was observed from the lamina propria to the epithelium in all samples taken from the marginal zone of peptic ulcers with or without *H. pylori* infection by the in situ fluorescence TUNEL method (Fig. 1). Reverse transcription–polymerase chain reaction findings showed the expression of perforin and Fas in all samples with and without *H. pylori* infec-

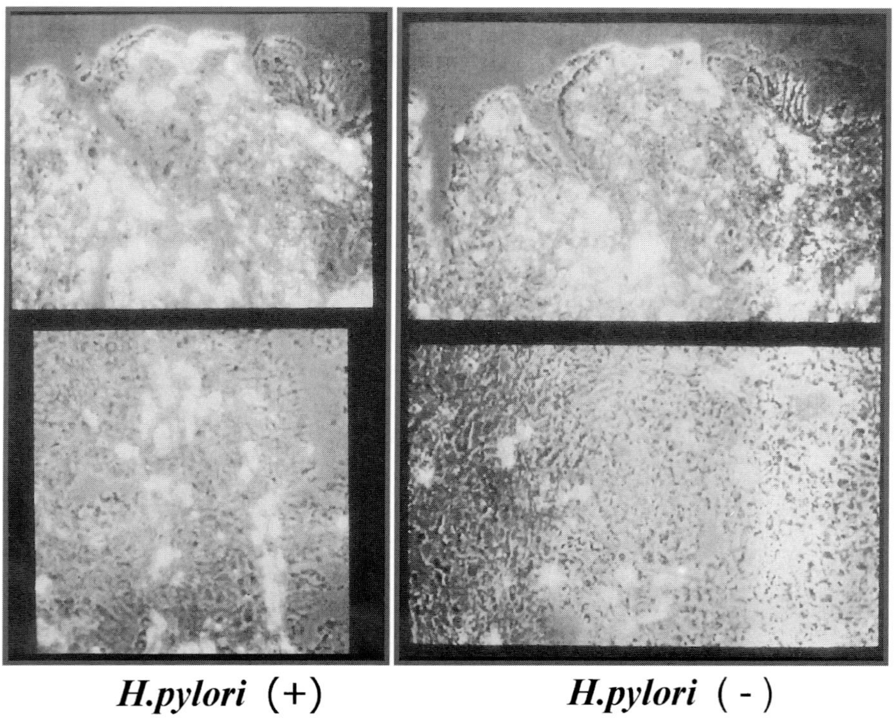

Fig. 1. Expression of apoptotic cells in the marginal zone of gastric ulcer by in situ fluorescence tunel method

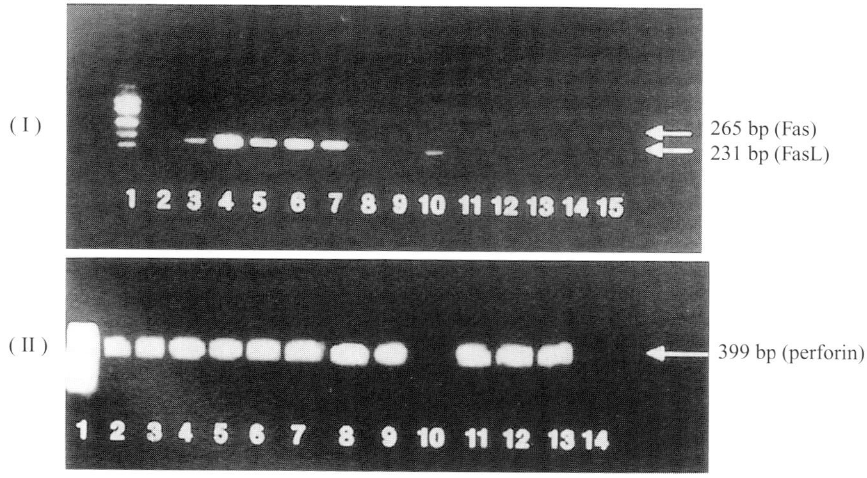

FIG. 2. Expression of Fas, FasL, and perforin in gastric and duodenal mucosa infected with or without *Helicobacter pylori* (HP) reverse transcription–polymerase chain rearta by RT-PCR

tion; however, FasL expression was never seen constantly (Fig. 2). Immunoblotting analysis showed constantly expression of caspase 3 in all samples regardless of *H. pylori* infection (Fig. 3). Immunohistochemical analysis revealed that perforin and granzyme B localized in the lesion from the CTL cytoplasm, between the proper mucous membrane and epithelium, to the target cell (Fig. 4). Moreover, caspase 3 was seen in granzyme B-expressing CTL and its surrounding interstitium (Fig. 5). Immunoelectron microscopy provided images of perforin released from cytoplasmic azurophile granules of CTL towards the target cells (Fig. 6a), and images of the target cells showed an early apoptotic appearance with a tight spherical body change and oscula in the cell membrane (Fig. 6b). There were also images of granzyme B that invaded the target cells through oscula to induce apoptosis of blood cells in cytoplasm, and in turn other cells, causing exfoliation (Fig. 7a,b).

Discussion

There are many reports on the relation between peptic ulcers and apoptosis, but reports on the pathway of apoptosis involved in peptic ulcers are limited [4]. Investigation of the effector of mucosal injury has never yet been reported except for our study [5]. By all means, the present clinical investigation alone

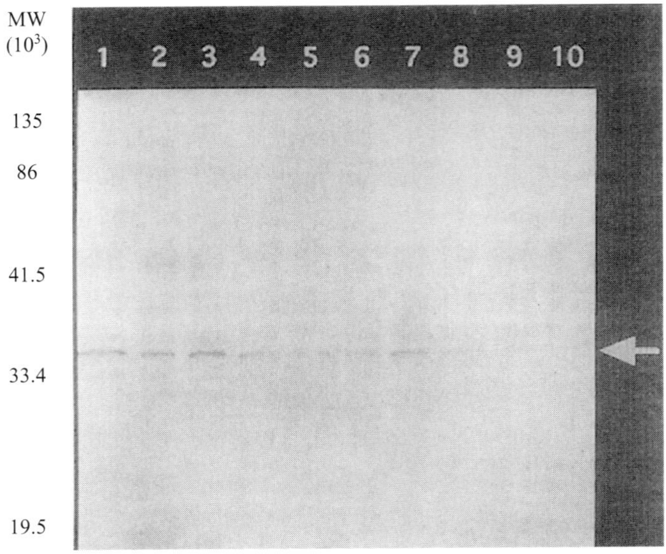

Fig. 3. Examination of caspase 3 by immunoblotting in gastric and duodenal mucosa with or without *H. pylori* infection. *MW*, molecular weight

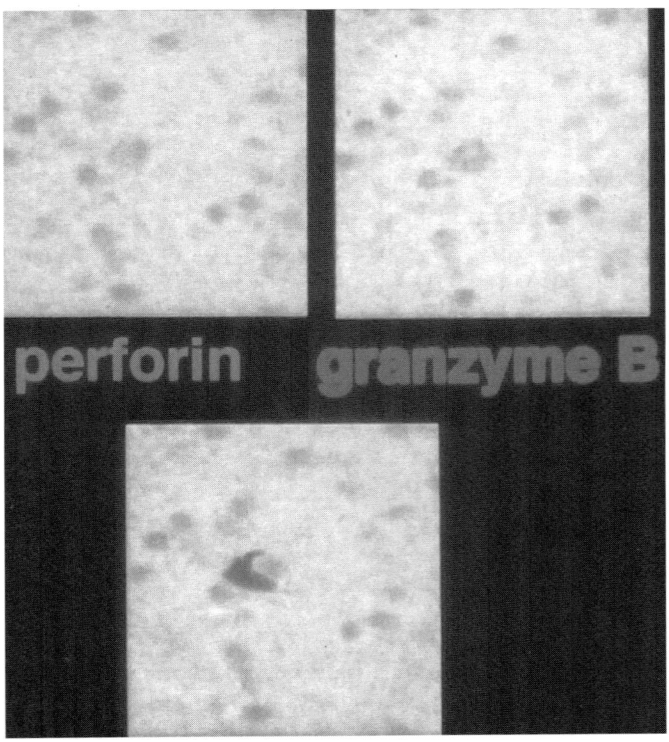

Fig. 4. Expression of perforin granzyme B by immunohistochemistry in the marginal zone of gastric ulcer

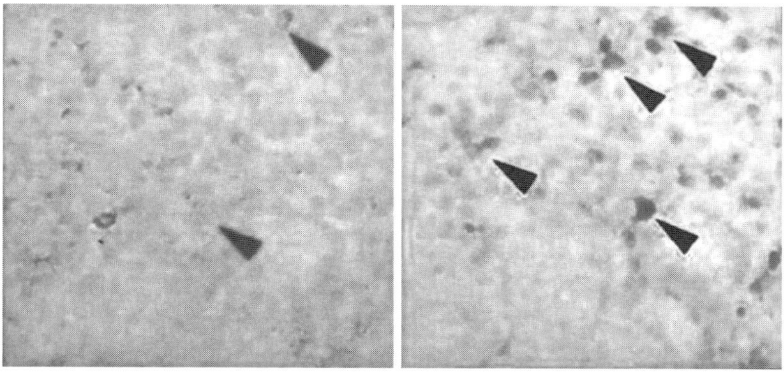

granzyme B caspase 3

Fig. 5. Expression of granzyme B caspase 3 by immunohistochemistry in the marginal zone of gastric ulcer

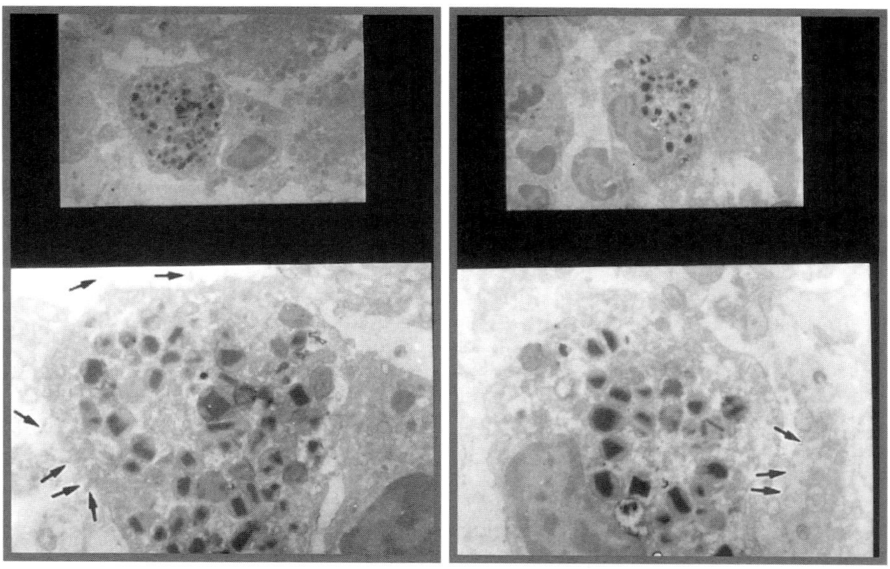

a b

Fig. 6a,b. Electron microscopic findings of perforin, which attacks the target cell by immunohistochemistry

cannot determine whether the observed phenomena may mean onset of mucosal damage or the course of regeneration. However, interestingly enough, it was suggested that CTL might be an effector (trigger) for mucosal damage irrespective of *H. pylori* infection. If it could be demonstrated that a certain substance might induce CTL with a common epitope in both groups irrespective of *H. pylori* infection, it would signify the possibility that there

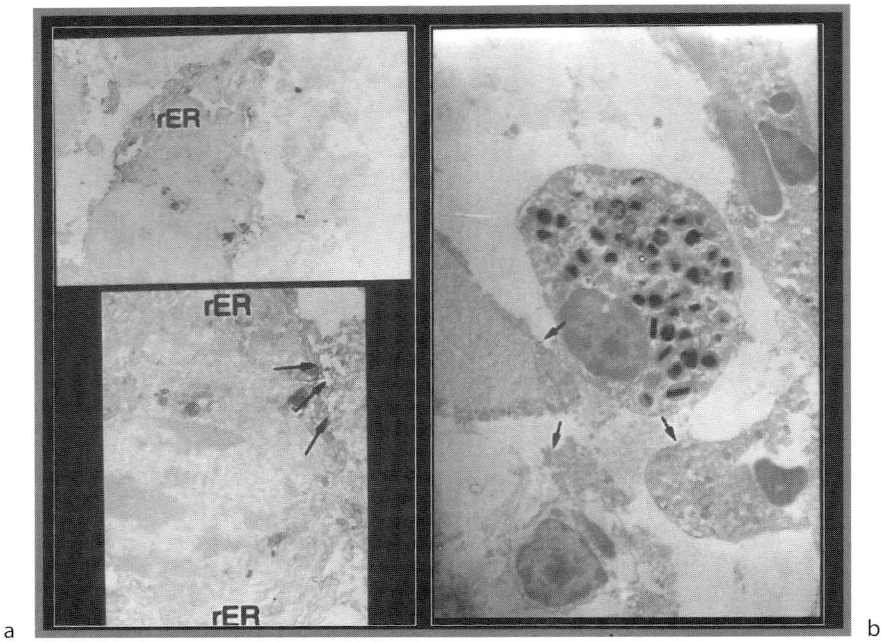

FIG 7a,b. Electron microscopic findings of granzyme B, which attacks the target cell by immunohistochemistry. ER, rough endoplasmic reticular

may be an unknown antigen responsible for peptic ulcers. If such a substance could be determined, it may be a significant discovery in terms of investigation of peptic ulcers, and further development can be expected. Whether we studied CTL derived from CD4 or CD8 using a flow cytometry in mononuclear cells obtained from mucosal biopsy specimens endoscopically, we could analyze it only for very few mononuclear cells. We have a schedule of examination concerning this unsolved problem using other methods. Thus we are planning further investigation of these various problems and questions using an experimental animal model and other clinical examinations based on these results.

References

1. Ando K, Hiroishi K, Kaneko T, et al (1997) Perforin, Fas/Fas ligand, and TNF-alpha pathways as specific and bystander killing mechanisms of hepatitis C virus-specific human CTL. J Immunol 158:5283–5291
2. Hattori K, Hirano T, Miyajima H, et al (1999) Differential effects of anti-Fas ligand and anti-tumor necrosis factor alpha antibodies on acute graft-versus-host disease pathologies. Blood 93:2738–2747

3. Sabelko Downers KA, Cross AH, Russell JH (1999) Dual role for Fas ligand in the initiation of and recovery from experimental allergic encephalomyelitis. J Exp Med 189: 1195–1205
4. Houghton JR, Korah RM, Condon M, et al (1999) Apoptosis and *Helicobacter pylori*-associated gastric and duodenal ulcer disease is mediated via the Fas antigen pathway. Dig Dis Sci 44:465–479
5. Ohara T, Morishita T, Suzuki H, et al (2003) Perforin and granzyme B of Cytotoxic T lymphocyte mediates apoptosis irrespective of *Helicobacter pylori* infection: Possible act as a trigger of peptic ulcer formation. Hepato-Gastroenterology 50:1774–1779

Re-Evaluation of Microcirculatory Endothelial Cell as an Endocrine System of Leptin: Its Significance During Ulcer Healing and *Helicobacter pylori* Infection

Masahiko Nakamura[1], Yasutada Akiba[2], Hidenori Matsui[3], Naoki Kumagai[4], Hiroshi Serizawa[4], Noriaki Watanabe[4], Satoshi Tsunematsu[4], Fumihiko Kaneko[4], Kouichiro Atsuta[1], Hiromasa Ishii[2], and Kanji Tsuchimoto[1,4]

Key words. Leptin, *Helicobacter pylori*, Acetic acid-induced ulcer, Endothelial cell, Myofibroblast

Introduction

The myofibroblast, i.e., activated or modified fibroblast, has been shown to have growth-promoting effects on endothelial cells and be closely associated to angiogenesis [1]. Our recent histochemical studies have revealed a marked increase in the number and size of myofibroblasts in the *Helicobacter pylori* (Hp)-infected human and Mongolian gerbil fundic mucosa [2,3]. On the other hand, the mediators which stimulate the conversion of undifferentiated mesenchymal cells or fibroblasts to the myofibroblasts remain to be clarified. Recently, leptin, a protein product of obese gene expressed primarily by adipocytes, has been shown to be related to angiogenesis [4], and the relation of leptin and myofibroblast is an interesting point, because myofibroblast derives from fat-storing or Ito cells. In this chapter, we discuss the immuno-histochemical changes of leptin and its Ob receptors during the healing of

[1] Department of Clinical Pharmacy and Clinical Sciences, School of Pharmaceutical Sciences, Kitasato University, 5-9-1 Shirokane, Minato-ku, Tokyo 108-8641, Japan
[2] Department of Internal Medicine, Keio University School of Medicine, 35 Shinanomachi, Shinjuku-ku, Tokyo 160-8582, Japan
[3] Kitasato Institute for Life Sciences, Kitasato University, 5-9-1 Shirokane, Minato-ku, Tokyo 108-8641, Japan
[4] Department of Internal Medicine, Kitasato Institute Hospital, 5-9-1 Shirokane, Minato-ku, Tokyo 108-8642, Japan

acetic acid-induced ulcer in rats and Hp-induced gastric mucosal damage in Mongolian gerbils.

Alteration of Leptin and Ob Receptors During Healing of Acetic Acid-Induced Ulcer

Stomach tissues were treated with Zamboni's fixative, and the indirect immunofluorescence method using monoclonal antibodies against leptin (YC-040, Yanaihara Institute, Shizuoka, Japan) was performed, counterstained with Alexa Fluor 594 phalloidin (Molecular Probes, Leiden, the Netherlands), and observed by confocal laser microscopy (Leica TCS NT). In the control group, leptin immunoreactivity was recognized in the mesenchymal cells in the lamina propria mucosa, i.e., endothelial cells of the microcirculatory network and fibroblasts showing weak alexa phalloidin reactivity. Some of the surface epithelial cells showed very weak leptin immunoreactivity (Fig. 1). Ob-R immunoreactivity was recognized in the mesenchymal cells surrounding the microvascular network as well as in the endothelial cells.

In the acetic acid-treated group, 100% acetic acid was attached to the serosal surface of the stomach for 30 s 7 days before the experiments. One week after the acetic acid-induced gastric ulcer formation, leptin receptor immunoreactivity significantly increased in the endothelial cells and surrounding fibroblasts. In addition, Ob-R immunoreactivity was shown to accumulate in the tip portion of the regenerated fundic glandular mesenchymal cells.

As to the leptin localization in the microcirculation, a relationship to angiogenesis has recently been pointed out, because leptin administration has been reported to bring about new fenestrated blood vessels [4]. Leptin has also been shown to synergistically stimulate angiogenesis with basic fibroblast growth factor and vascular endothelial growth factor, the two most potent and commonly expressed angiogenic factors. As a whole, leptin has been suggested to be closely linked to the regeneration of the gastric mucosa after acetic acid-induced gastric mucosal injury.

Alteration of Leptin and Ob Receptors in Hp-Induced Gastric Mucosal Damage

Helicobacter pylori infection was evoked in Mongolian gerbils by the oral administration of CagA- and VacA-positive Hp strain (ATCC 43504) 12 months before the experiments. In this group leptin was markedly recognized in the mesenchymal cells including myofibroblasts which showed strong

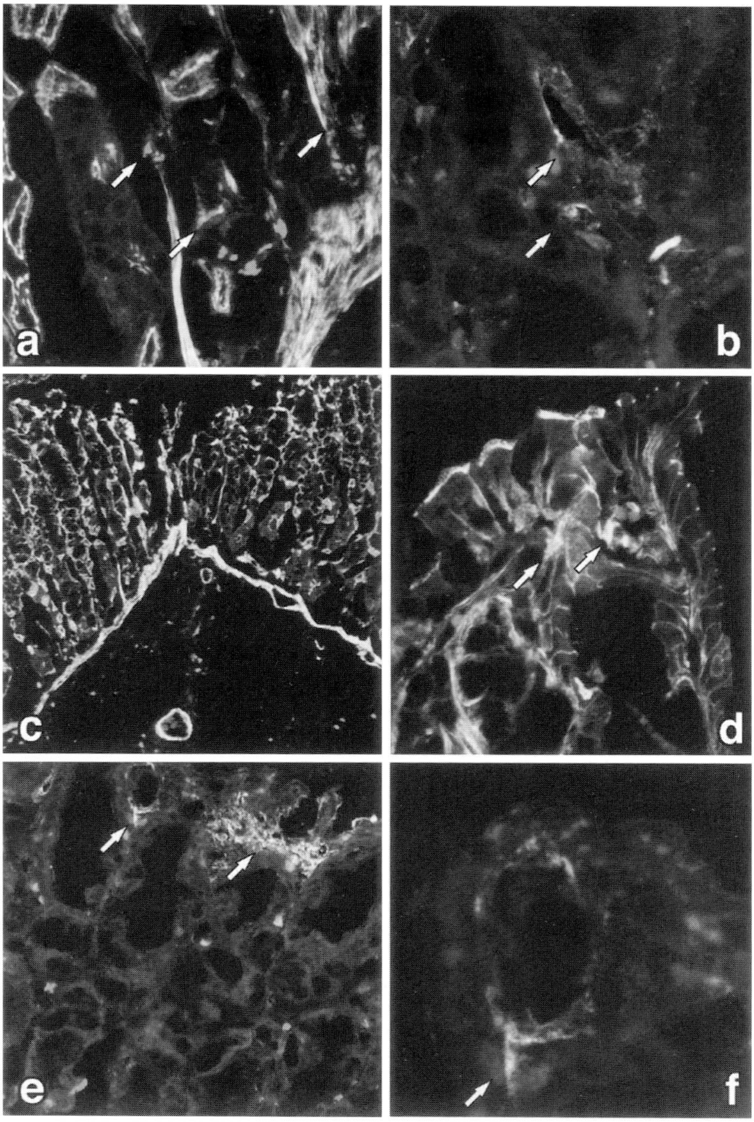

FIG. 1a–f. Confocal laser microscopic observation of leptin and Ob-R immunoreactivities in control and acetic acid-treated rat fundic mucosa. **a** In the base of the fundic mucosa, many endothelial cells (*arrows*) show a positive reaction to leptin. **b** In the tip of the fundic mucosa, many mesenchymal cells (*arrows*) surrounding the microvascular network are immunoreactive to Ob-R. **c,d** In the acetic acid-treated gastric mucosa, erosive lesion is observed (**c**). In the regenerating mucosa, many enlarged endothelial cells are reactive to leptin (*arrows*). **e,f** In the regenerating mucosa, strong Ob-R immunoreactivity is recognized in the mesenchymal cells (*arrows*)

alexa-phalloidin reactivity and endothelial cells (Fig. 2) [5]. Ob-R immunoreactivity accumulated markedly in the myofibroblasts, which were identified by myofibroblast-specific PR 2D3 immunoreactivity [6]. We have reported a markedly increased number of myofibroblasts during acetic acid-induced ulcer healing in rats [2] and Hp infection in gerbils [3]. Our result suggests the interaction of leptin with this transformation of fibroblast to myofibroblast.

Alteration of Leptin and Ob Receptors and Transformation of Fibroblast in Two-Layered Culture Model

In the in vitro experiment, GSM06 cells, i.e., gastric surface mucous cell line from a primary culture of gastric fundic mucosal cells of transgenic mice, were incubated in several media with an air–liquid surface on the collagen gel layer containing transformed mouse fibroblast cell line L929 for 2 weeks before experiments. The medium containing Hp bacilli (ATCC 43504) at multiplicity of infection of 10–100 were added from the luminal side and incubated with 5% CO_2 for 1–24h. *Helicobacter pylori* was found to attach to the GSM06 cells with pedestal formation and the number of the microvilli was increased on the surface plasma membrane of these mucous cells (Fig. 3). L929 cells in the collagen gel showed a spindle shape with poor cytoplasm. In the Hp-treated group, some of the L929 cells under GSM06 cells showed immunoreactivity of leptin and Ob-RL as well as PR 2D3. On electron microscopic observation, these cells had indented nuclei, well-developed r-ER, and actin bundles with dense patches just under plasma membrane, characteristic of the myofibroblast. The L929 cells incubated with Hp alone could not show differentiation to myofibroblasts. Pretreatment with leptin antibody prevented the transformation of the fibroblast. These data support the idea that leptin may stimulate fibroblasts to differentiate into myofibroblasts. Many cytokines play a role in the regulation of fibroblast activities, including growth factors, interleukins, and retinoic acid. The inter-relationship between leptin and these factors requires clarification.

Conclusion

Fibroblasts can be transformed to myofibroblasts during the healing process and Hp infection at least, in part, through the effect of leptin from the endothelial cells.

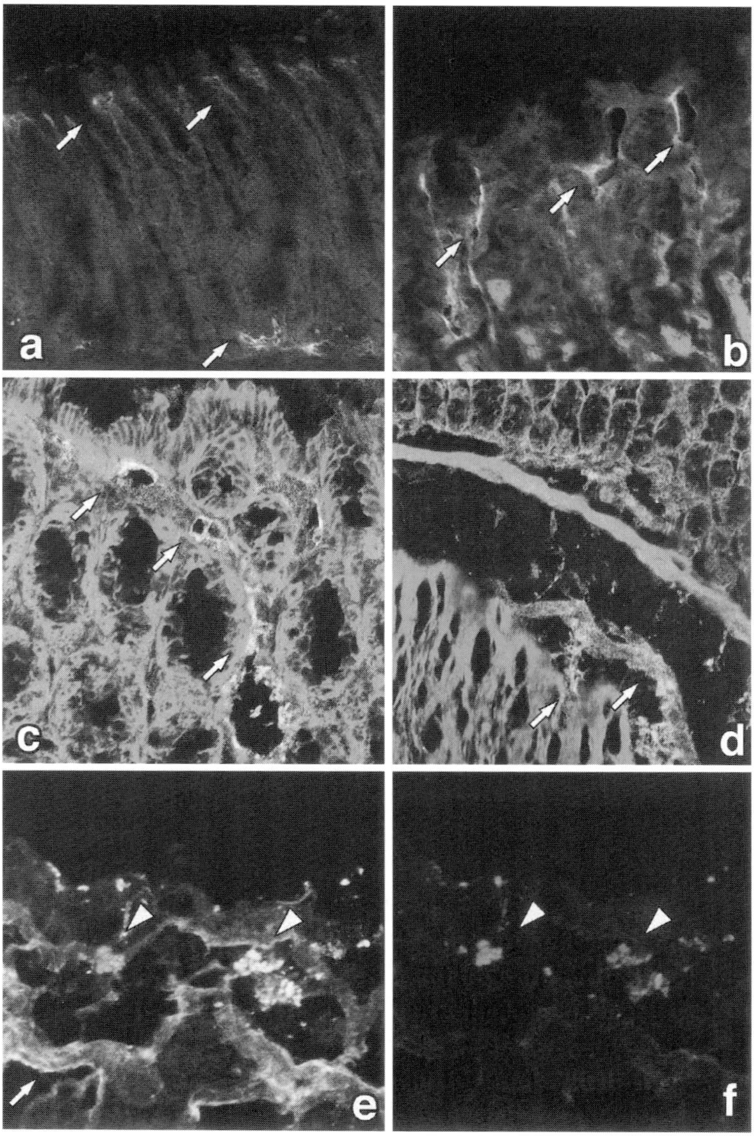

FIG. 2a–f. Confocal laser microscopic observation of leptin and Ob-R immunoreactivities in control and Hp-infected Mongolian gerbil fundic mucosa. **a,b** In the control Mongolian gerbil fundic mucosa, leptin immunoreactivity is seen in the endothelial cells of the collecting venules in the tip portion and the metarterioles in the base of the mucosal layer (*arrows*). **c,d** In the Hp-infected Mongolian gerbil fundic mucosa, the cytoplasm of the endothelial cells has enlarged and is positive to leptin (*arrows*). **e,f** Ob-R immunoreactivity is found in the endothelial cells and mesenchymal cells in the tip portion of the fundic mucosa of the Hp-infected Mongolian gerbils (*arrowheads*)

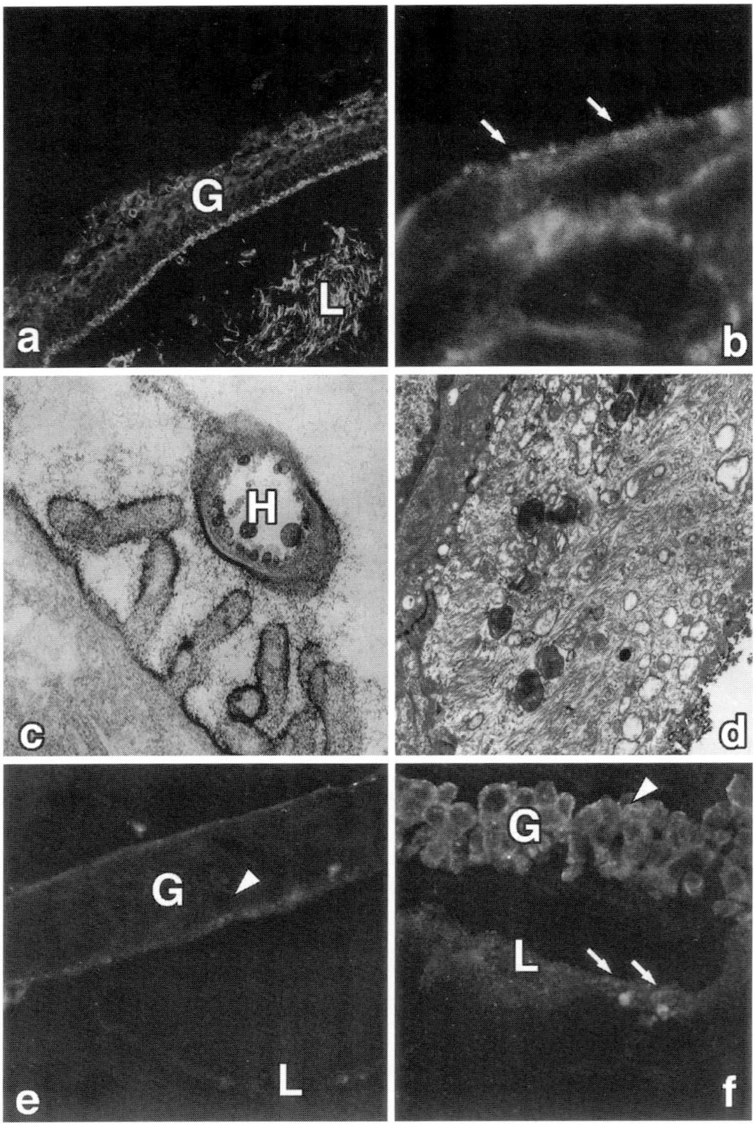

FIG. 3a–f. Confocal laser and electron microscopic observation of the GSM06 cells and L929 cells. **a,b** The culture cells are composed of the GSM06 cells (*G*) and L929 cells (*L*). Attached Hp bacilli are seen in the luminal surface of the GSM06 cells (*arrows*). **c** Electron microscopic observation of Hp (*H*) almost attached to the GSM06 cells. **d** Many microfilaments are recognized in the L929 cells after Hp attachment to the GSM06 cells. **e** In the control group, leptin immunoreactivity is very weak in GSM06 cell layers (*arrowhead*). **f** In the Hp-treated group, leptin immunoreactivity in the GSM06 cells (*arrowhead*) and Ob-R reactivity in the L929 cells (*arrows*) are clearly recognized

References

1. Nicosia RF, Tuszynski GP (1994) Matrix-bound thrombospondin promotes angiogenesis in vitro. J Cell Biol 124:183–193
2. Nakamura M, Oda M, Inoue J, et al (1995) Plasticity of myofibroblasts appearing in granulation tissues after acetic acid treatment effect of bFGF. Dig Dis Sci 40:2477–2480
3. Nakamura M, Akiba Y, Oda M, et al (1997) Alteration of basic fibroblast growth factor concentration and immunoreactivity in healing of ethanol-induced gastric mucosal damage: effect of sofalcone. J Clin Gastroenterol 25(suppl 1):S13–S20
4. Cao RBE, Brakenhielm E, Wahlestedt C, et al (2001) Leptin induces vascular permeability and synergistically stimulates angiogenesis with FGF-2 and VEGF. Proc Natl Acad Sci USA 98:6390–6395
5. Nakamura M, Akiba Y, Matsui H, et al (2003) Interaction of leptin with gastric myofibroblast transdifferentiation in Hp-infected Mongolian gerbils: effect of rebamipide. Aliment Pharmacol Ther 18(suppl 1):99–105
6. Richman PI, Tilly R, Jass JR, et al (1987) Colonic pericrypt sheath cells: characterization of cell type with new monoclonal antibody. J Clin Pathol 40:593–600

Bile Nucleotides Exaggerate Ischemia–Reperfusion-Induced Epithelial Injury via P2Y, Not P2X Purinoceptor in Rat Jejunum

Misa Mizumori[1], Yasutada Akiba[1,2], Soichiro Miura[3], Hidekazu Suzuki[1], Hiroshi Nagata[1], and Hiromasa Ishii[1]

Key words. P2 receptor, Bile acid, Ischemia–reperfusion, Jejunal epithelial injury, Intracellular pH

Introduction

Extracellular nucleotides have important roles to mediate and regulate the physiological functions such as neurotransmission and secretion [1,2]. Bile contains enough concentration of nucleotides to activate purinoceptors [3], such as P2X and P2Y adenosine triphosphate (ATP) receptor expressed in the lumen of epithelial cells [2]. However, the effects of luminal nucleotides on the intestinal epithelial functions are unknown. We have demonstrated that luminal taurocholic acid (TCA) acidifies epithelial cells and enhances ischemia–reperfusion (I/R)-induced epithelial injury in rat jejunum [4]. The present study was undertaken to examine the effects of luminal nucleotides with or without TCA on intracellular pH (pH_i) and cell injury induced by I/R in rat jejunal epithelia.

[1] Department of Internal Medicine, Keio University School of Medicine, 35 Shinanomachi, Shinjuku-ku, Tokyo 160-8582, Japan
[2] CURE/UCLA & BBRI, Building 114, Suites 217, WLA VAMC, 11301 Wilshire Boulevard, Los Angeles, CA 90073, USA
[3] Second Department of Internal Medicine, National Defense Medical College, 3-2 Namiki, Tokorozawa-shi, Saitama 359-8513, Japan

Materials and Methods

Male Wistar rats (250–300 g) were used. Under urethane anesthesia (1.25 g/kg i.p.), the abdomen was opened and the posterior wall of the proximal jejunum 2 cm distal to the ligament of Treitz was exposed. Using in vivo fluorescent microscopy, pH_i was measured with BCECF loaded into the epithelial cells, and epithelial injury was assessed by propidium iodide (PI) staining in vivo in situ as previously described for rat duodenum [5]. After the stabilization and time was set as $t = 0$, the jejunal mucosa was topically superfused through the mucosally-placed chamber with Krebs buffer (pH 7.4) for 5 min, followed by Krebs superfusion with or without TCA (20 mM), P2X agonist αβ-methylene-ATP (αβMeATP, 0.1 mM), P2X antagonist pyridoxal phosphate-6-azo (benzene-2,4-disulfonic acid) (PPADS, 0.1 mM), P2Y agonist 2-methylthio-ATP (2MeSATP, 0.1 mM), P2X/P2Y antagonist suramin (0.1 mM), or P2Y2 and P2Y4 agonist UTP (0.1 mM) (from $t = 5$ min to $t = 35$ min). Ischemia–reperfusion was performed by a 5-min occlusion of the anterior mesenteric artery (ischemia, from $t = 5$ to $t = 10$) followed by the release of the occlusion (reperfusion, from $t = 10$ to $t = 35$).

Results

αβ-Methylene-ATP and 2MeSATP acidified the epithelial cells, whereas PPADS and suramin alkalinized cells. Each chemical alone had no effect on PI-positive cell number. pH_i was transiently decreased by 5 min I/R without cell injury. Under this "mild" I/R, 2MeSATP enhanced cell injury, whereas others did not. Taurocholic acid alone acidified cells without injury, whereas TCA under mild I/R further acidified cells and induced cell injury. 2-Methylthio-ATP exaggerated and suramin abolished TCA+I/R-induced injury. Furthermore, UTP enhanced TCA+I/R-induced injury by a suramin-inhibitable mechanism.

Discussion

We demonstrated that the cellular acidification by luminal TCA, luminal P2 agonists, and mild ischemia potentiates triggering the epithelial injury in rat jejunum. Furthermore, P2Y, not P2X receptor, exaggerates the TCA toxicity and I/R-induced epithelial injury. Suramin, a nonspecific P2 receptor antagonist, alkalinized cells, consistent with the observation in the bronchial epithelium [6], and 2MeSATP, a P2Y receptor agonist, enhanced, but suramin inhibited, the TCA-induced acidification, suggesting that P2Y receptor is spontaneously activated and regulates the TCA absorption under physiological condition presumably with bicarbonate secretion, since P2Y is involved in the bicarbonate secretion [7]. Furthermore, 2MeSATP exaggerated, but suramin reduced or abolished, the injurious effects of TCA+I/R and UTP, suggesting that P2Y (P2Y2 or P2Y4) activation is involved in the epithelial injury in the ischemic intestine. These findings may account for the chain injurious reaction of epithelial injury during the ischemia due to the leaked extracellular ATP. P2Y antagonist may be useful to reduce the epithelial injury in the ischemic intestine, e.g., surgical operation and small intestine transportation.

References

1. Roman RM, Fitz JG (1999) Emerging roles of purinergic signaling in gastrointestinal epithelial secretion and hepatobiliary function. Gastroenterology 116:964–979
2. Leipziger J (2003) Control of epithelial transport via luminal P2 receptors. Am J Physiol 284:F419–F432
3. Chari RS, Schutz SM, Haebig JE, et al (1996) Adenosine nucleotides in bile. Am J Physiol 270:G246–G252
4. Mizumori M, Akiba Y, Miura S, et al (2002) Taurocholate, not tauroursodeoxycholate enhances ischemia-reperfusion-induced epithelial injury in rat jejunum (abstract). Gastroenterology 122:M1151
5. Akiba Y, Furukawa O, Guth PH, et al (2001) Cellular bicarbonate protects rat duodenal mucosa from acid-induced injury. J Clin Invest 108:1807–1816
6. Urbach V, Helix N, Renaudon B, et al (2002) Cellular mechanisms for apical ATP effects on intracellular pH in human bronchial epithelium. J Physiol 543:13–21
7. Dranoff JA, Masyuk AI, Kruglov EA, et al (2001) Polarized expression and function of P2Y ATP receptors in rat bile duct epithelia. Am J Physiol 281:G1059–G1067

Prostaglandin E1 and a Serine Protease Inhibitor Protect the Gastric Microcirculation and Increase the Gastric Acid Secretion After Thermal Injury

Tetsuya Nakamura[1], Masashi Yoshida[1], Hideki Ishikawa[2], Go Wakabayashi[1], Motohide Shimazu[1], Minoru Tanabe[1], Shigeyuki Kawachi[1], Koichiro Kumai[3], Tetsuro Kubota[1], Yoshihide Otani[1], Yoshiro Saikawa[1], and Masaki Kitajima[1]

Key words. Prostaglandin E1, Serine protease inhibitor, Gastric acid, Gastric microcirculation

Microcirculatory disturbance of the gastric wall is a crucial factor in the development of gastric mucosal lesions induced by *Helicobacter pylori* [1–3], anti-inflammatory drugs [4,5], and stress [6,7]. After thermal injury to the dorsal skin in rats, macroscopic hemorrhagic erosion developed in 14.3% at 15 min, 42.9% at 2 h, 100% at 5 h, and 85.7% at 12 h [8]. Macroscopic hemorrhagic erosion is formed within 5 h after thermal injury. Under the stereoscopic microscope, superficial gastric erosion could be observed in all rats studied at 15 min. Gastric mucosal blood flow decreased at 15 min, partially improved at 2 h, and decreased again at 5 h after thermal injury [9]. Since the blood flow was depressed especially at 15 min and 5 h, we have investigated mechanisms of the decrease in gastric mucosal blood flow at 15 min and 5 h. Silicon rubber casts of the vasculature revealed contraction of arterioles 15 min after thermal injury to the dorsal skin in the rat model [10]. We also observed the gastric microvessels by in vivo microscopy which was first described by Guth and

[1] Department of Surgery, Keio University School of Medicine, 35 Shinanomachi, Shinjuku-ku, Tokyo 160-8582, Japan
[2] Department of Emergency and Critical Care Medicine, Keio University School of Medicine, 35 Shinanomachi, Shinjuku-ku, Tokyo 160-8582, Japan
[3] Center for Diagnostic and Therapeutic Endoscopy, Keio University School of Medicine, 35 Shinanomachi, Shinjuku-ku, Tokyo 160-8582, Japan

Rosenberg [11,12]. In vivo microscopic observation of gastric microcirculation showed also the contraction of arterioles 15 min after thermal injury [13]. The contraction can be responsible for the decrease in gastric mucosal blood flow 15 min after thermal injury. Five hours after thermal injury, in vivo microscopy showed irregular constriction of the venules caused by suppressed production of nitric oxide [14]. The addition of leukocytes stimulated by phorbol myristate acetate to the endothelial cell monolayer caused a significant increase in the intracellular peroxide level in the endothelial cells and severe endothelial cell injury after 5 h, thus suggesting the presence of leukocyte-dependent endothelial damage [15].

Rats in the prostaglandin E1 (PGE1) group were administered 1.0 μg/kg per minute intravenously from 30 min before thermal injury. Microvascular images in the basal region of the gastric mucosa were observed by in vivo microscopy, and carboxyfluorescein diacetate succinimidyl ester dependent leukocyte illumination was monitored. Monastral blue B (MBB) was administered intravenously and deposits of MBB in venules were observed. Diameters of arterioles and accompanied venules were measured and arteriolar diameter/venular diameter (A/V ratio) was calculated at three different points. The gastric effluent was perfused (10 ml/h), and was collected every 1 h. The percentage of rolling leukocytes was significantly higher in rats at 2 h after injury than that in the normal control group, and significantly lower in the PGE1 group than that in the saline-administered group. Percentage of the MBB deposits area was significantly higher in the 2 h after injury group than that in the normal control group, and significantly lower in the PGE1 group than that in the saline-administered group. A/V ratio in the 2 h group was significantly smaller than that in the normal control group. In the PGE1 group, gastric mucosal lesion was suppressed, and rolling of leukocyte and area of MBB deposits were inhibited. One hour after thermal injury, gastric acid secretion was increased in the PGE1 group.

A serine protease inhibitor, camostat mesilate (CM) (100 mg/kg), was administered i.g. 30 min before thermal injury. In the CM group, gastric mucosal lesion was suppressed, and rolling of leukocyte and area of MBB deposits were inhibited. Before thermal injury, CM did not affect gastric acid secretion. After thermal injury, gastric acid secretion was decreased in the distilled water-administered rats and was increased in the CM group (Fig. 1).

Gastric acid secretion is increased in animal experimental models of stress. However, a rapid decrease in the gastric mucosal blood flow, energy charge depression, damage to the mitochondria in parietal cells, parietal resting stage, and low acid output were observed after thermal injury [16]. It was hypothesized that microcirculatory disturbance was the cause of the low acid output after thermal injury. It is suggested that CM and PGE1 inhibit gastric mucosal lesion at least in part by protection against microcirculatory distur-

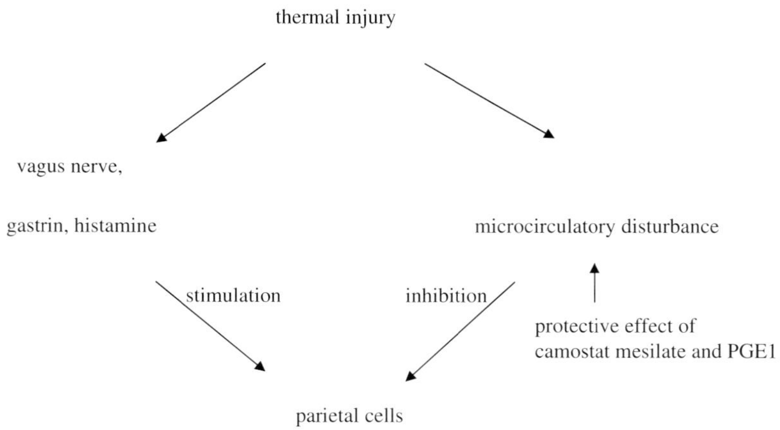

FIG. 1. The possible mechanisms of acid secretion under stress. *PGE1*, prostaglandin E1

bance. After thermal injury, inhibition of gastric acid secretion caused by microcirculatory disturbance may be superior to the acid-secretory stimulation. Thus, protection of the microcirculation by CM and PGE1 may result in an increase in acid output.

References

1. Kalia N, Jones C, Bardhan DK, Reed MW, Atherton JC, Brown NJ (2001) Effects of genotypically different strains of Helicobacter pylori on human microvascular endothelial cells in vitro. Dig Dis Sci 46:54–61
2. Kalia N, Bardhan KD, Reed MV, Jacob S, Brown NJ (2000) Mechanisms of Helicobacter pylori-induced rat gastric mucosal microcirculatory disturbances in vivo. Dig Dis Sci 45:763–772
3. Kalia N, Bardhan KD, Reed MV, Jacob S, Brown NJ (2000) Effects of chronic administration of Helicobacter pylori extracts on rat gastric mucosal microcirculation in vivo. Dig Dis Sci 45:1343–1351
4. Kawano S, Tsuji S (2000) Role of mucosal blood flow: a conceptional review in gastric mucosal injury and protection. J Gastroenterol Hepatol 15(suppl):D1–D6
5. Fiorucci S, Antonelli E, Morelli O, Morelli A (1999) Pathogenesis of non-steroidal anti-inflammatory drug gastropathy. Ital J Gastroenterol Hepatol 31(suppl 1):S6–S13
6. Filatetova L, Maltcev N, Bogdanov A, Levkovich Y (1999) Role of gastric microcirculation in the gastroprotection by glucocrticoids released during water-restraint stress in rats. Chin J Physiol 42:145–152
7. Yoshida M, Wakabayashi G, Ishikawa H, Otani Y, Shimizu M, Kubota T, Kumai K, Kurose I, Miura S, Ishii H, Kitajima M (1998) A protease inhibitor attenuates gastric erosions and microcirculatory disturbance in the early period after thermal injury in rats. J Gastroenterol Hepatol 13:104–108
8. Yoshida M, Kitahora T, Wakabayashi G, Tashiro H, Ono H, Otani Y, Shimazu M, Kubota

T, Kumai K, Kitajima M (1995) Active oxygen species in formation of acute gastric mucosal lesions induced by thermal injury in rats. Dig Dis Sci 40:1306–1310
9. Yoshida M, Furukawa D, Wakabayashi G, Otani Y, Oshima A, Shimazu M, Kubota T, Kumai K, Kurose I, Miura S, et al (1995) Gastric microcirculatory disturbance and behavior of leucocytes after thermal injury: intravital observation of arteriovenous shunting channels in the gastric submucosal layer. J Gastroenterol Hepatol 10:365–370
10. Kitajima M, Otuka S, Shimizu A, Nakajima M, Kiuchi T, Ikeda Y, Oshima A (1998) Impairment of gastric microcirculation in stress. J Clin Gastroenterol 10:S120–S128
11. Rosenberg A, Guth PH (1970) A method for the in vivo study of the gastric microcirculation. Microvasc Res 2:111–112
12. Guth PH, Rosenberg A (1972) In vivo microscopy of the gastric microcirculation. Am J Dig Dis 17:392–398
13. Yoshida M, Wakabayashi G, Otani Y, Oshima A, Shimazu M, Kubota T, Kumai K, Kurose I, Miura S, Kitajima M (1995) Active oxygen species generation by circulating leukocytes and gastric submucosal microcirculatory disturbances in the early period after thermal injury. J Clin Gastroenterol 21(suppl 1):S87–S92
14. Yoshida M, Kurose I, Wakabayashi G, Hokari R, Ishikawa H, Otani Y, Shimazu M, Miura S, Ishii H, Kitajima M (1997) Suppressed production of nitric oxide as a cause of irregular constriction of gastric venules induced by thermal injury in rats. J Clin Gastroenterol 25(suppl 1):S56–S60
15. Murota SI, Fujita H, Morita I, Wakabayashi Y (1995) Adhesion molecules mediated endothelial cell injury elicited by activated leukocytes. Ann NY Acad Sci 748:133–147
16. Kitajima M (1990) Recent investigation and discussion about cytoprotection. Prog Med 10:2211–2215

Duodenal Blood Flow in Acute Portal Hypertension

Eisuke Iwasaki[1], Tetsuo Morishita[1], Eiichi Sekizuka[2], Kouji Miyazaki[2], Takashi Osada[1], Hiroshi Kishikawa[1], Masaru Nakano[1], Hidekazu Suzuki[4], Tadashi Ohara[1,3], Jiro Nishida[1], Hiroshi Nagata[4], and Hiromasa Ishii[4]

Key words. Microcirculation, Duodenum, Portal hypertension, In vivo microscopy, Blood vessels

Introduction

We have already reported gastric microcirculatory changes in portal hypertension [1]. Duodenal mucosal lesions such as erosions are often seen in patients with portal hypertension. The purpose of this study was to examine the changes of the mucosal and submucosal vessels in the rat duodenum during the stepwise increase of the portal vein pressure.

Materials and Methods

Male Wister rats fasted overnight were anesthetized with 50 mg/kg sodium pentobarbital i.p. A fine vinyl tube and polyethylene tube were utilized to make a small circle around the portal vein at the porta hepatis for partial ligation of the portal vein. The pressure was increased in a stepwise manner from the control up to +5 cmH_2O, +10 cmH_2O, +15 cmH_2O, and +20 cmH_2O. The aimed pressure (±10%) was maintained for 2 min at each level. Duodenal blood flow, 1–3 cm anal to the pyloric ring, was measured from the mucosal side with a laser-Doppler flowmeter (ALF 2100, Advance, Japan). An in vivo

[1] Departments of Internal Medicine and Gastroenterology, Tokyo Dental College, Ichikawa General Hospital, 5-1-13 Sugano, Ichikawa, Chiba 272-8513, Japan
[2] Saitama National Hospital, 2 Suwa, Wako, Saitama 351-0102, Japan
[3] Tokyo Dental College, Chiba Hospital, 1-1-2 Masago, Mihama-ku, Chiba 261-8502, Japan
[4] Department of Internal Medicine, Keio University School of Medicine, 35 Shinanomachi, Shinjuku-ku, Tokyo 160-8582, Japan

microscopy technique was used to study the mucosal and submucosal microvasculature changes. Blood cell velocity in the duodenal mucosal capillary was measured with the CapiFlow (Stockholm, Sweden) system. Diameter of the submucosal arterioles and venules was measured with an image analysis system (Avionix TVIP-2000; NEC PC-9801, Japan). Mean duodenal blood flow and blood cell velocity in the capillaries were taken during the last 1 min at each level of the portal pressure.

Results

Duodenal blood flow decreased as the portal pressure increased from the control (49.0 ± 4.6 ml/min per 100 g, Mean ± SD) up to +20 cmH_2O (18.6 ± 1.8 ml/min per 100 g) with significant ($P < 0.05$) differences between the control and at +5 cmH_2O (40.0 ± 4.6 ml/min per 100 g), +5 cmH_2O and +10 cmH_2O (33.4 ± 4.0 ml/min per 100 mg), and +20 cmH_2O (Fig. 1). Blood cell velocity in the mucosal capillaries decreased from the control (0.12 ± 0.02 mm/s) with significant differences as the portal pressure increased.

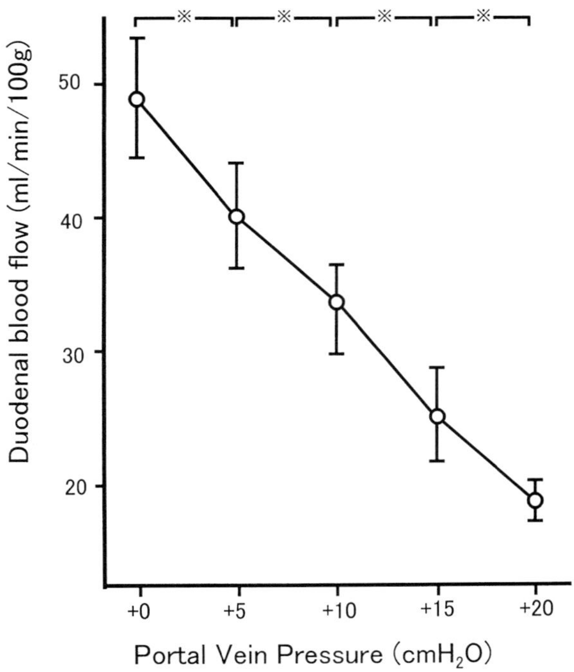

FIG. 1. Duodenal blood flow during the increase in portal vein pressure. Duodenal blood flow decreased from the control (49.0 ± 4.6 ml/min per 100 g, mean ± SD, seven rats) with significant (*$P < 0.05$) differences as the portal pressure increased

Diameter of the submucosal arterioles decreased from the control (26.7 ± 2.0 μm) with significant differences as the portal pressure increased. Diameter of the submucosal venules (53.3 ± 3.1 μm) showed no significant differences among the levels.

Discussion

Acute portal hypertension, which is induced by inflammation, tumor, or thrombosis, is often accompanied with abdominal pain and fatal hemorrhagic necrosis of the intestine. Our study showed that stepwise increase of the portal vein pressure decreased the blood flow, the blood cell velocity of the mucosal capillaries, and the diameter of submucosal arterioles in the duodenum. This phenomenon seems to be similar to gastric and ileal microcirculatory changes [1,2]. The mechanism of the duodenal microcirculatory change is still unknown. In the stomach, both endothelin and nitric oxide are involved in the diameter changes of submucosal arterioles, and the arterioles are constricted through the predominant function of endothelin over nitric oxide [2,3]. Further studies on the mechanism of duodenal microcirculatory changes are needed. In conclusion, duodenal blood flow decreases with the constriction of submucosal arterioles in acute portal hypertension.

References

1. Morisita T, Iwasaki E, Ishii H, et al (1990) Microcirculatory changes of the stomach in the portal hypertension. Microcirc Annu 5:99–100
2. Nagata H, Sekizuka E, Morisita T, et al (1993) Regional differences in microvascular response in rat intestine during acute elevation of portal pressure. J Gastroenterol Hepatol 8:315–321
3. Morisita T, Iwasaki E, Ishii H, et al (1993) Role of endothelin and nitric oxide in gastric submucosal arteriolar changes in acute portal hypertension. Microcirc Annu 9:109–110

Part 3
Liver Microcirculation Research for Clinical Application

Genetic Manipulation of Liver Sinusoidal Endothelial Cells

Yoshiyuki Takei[1], Kenichi Ikejima[1], Nobuyuki Enomoto[1], Atsushi Maruyama[2], and Nobuhiro Sato[1]

Summary. The liver sinusoidal endothelial cells (SECs) possess unique hyaluronan receptors that recognize and internalize hyaluronic acid (HA). This characteristic was used in the development of a system for targeting foreign DNA to SECs. A gene carrier system was prepared by coupling HA (number-average molecular weight: 1.5×10^4) to poly-L-lysine (PLL, number-average molecular weight: 4.6×10^4) in a 1:1 weight ratio by reductive amination reaction. The resulting copolymer (PLL-g-HA) was mixed with various amounts of DNA in 154 mM NaCl at 4°C. Inter-polyelectrolyte complex formation between PLL-g-HA and DNA exhibited minimal self-aggregation, explaining the highly soluble nature of the complex. The agarose gel retardation assay revealed that the titration point representing the minimum proportion of PLL-g-HA required to retard the DNA completely occurred at a 1:1 copolymer to DNA charge ratio. Intravenous injection of the [^{32}P]pSV β-Gal plasmid complexed to PLL-g-HA in Wistar rats demonstrated specific hepatic targeting with >93% of the injected counts taken up by the liver in 1 h. Further, using a fluorescein isothiocyanate-labeled DNA, it was shown that the PLL-g-HA/DNA complex was distributed exclusively in the SECs. Seventy-two hours after injection of 90 μg of pSV β-Gal in a PLL-g-HA-complexed form, a large number of SECs expressing β-galactosidase were detected. Thus, the PLL-g-HA/DNA system permits targeted delivery of exogenous genes selectively to the liver SECs, providing a novel strategy for manipulation of SEC functions.

Key words. Sinusoidal endothelial cell, PLL-g-HA, Triplex DNA, Decoy DNA duplex

---

[1] Department of Gastroenterology, Juntendo University School of Medicine, 2-1-1 Hongo, Bunkyo-ku, Tokyo 113-8421, Japan
[2] Institute of Materials Chemistry and Engineering, Kyushu University, 6-10-1 Hakozaki, Higashi, Fukuoka 812-8581, Japan

Introduction

Liver sinusoidal endothelial cells (SECs) play a pivotal role in modulating immunity and inflammation [1,2]. Altered immunoresponses of SECs are closely related to pathogenesis of viral hepatitis, fulminant hepatic failure and transplant rejection [3]. Further, interest in xenotransplantation has grown because of the increased demand for donor organs and the possibility of avoiding post-transplant recurrence of viral hepatitis [4]. Xenograft livers usually undergo rapid rejection. It has been proposed that "activation of the SEC" underlies the vascular aspects of discordant xenotransplant rejection. The phenomena are characterized by rapid gene expression in SECs of procoagulant proteins and proinflammatory mediators [5].

Given the importance of altered gene expression in SECs, it is possible to envision a new therapeutic strategy for treatment of lethal liver diseases and achievement of graft-specific immunotolerance through modulation of SEC functions by genetic engineering. However, currently this approach is impossible because of lack of modality of selective delivery of nucleotide agents to SECs.

SECs possess unique receptors that recognize and internalize hyaluronic acid (HA) [6,7]. We utilized this characteristic of SECs in the development of a system for targeting foreign DNA to SECs.

Synthesis of PLL-g-HA Comb-Type Copolymers

High molecular weight HA (5.9×10^5) was partially degraded by hyaluronidase to low-molecular weight fragments. Hyaluronidase and HA was added together in 120 ml of water. The solution was stirred at 50°C for a desired time (1–20 h). Then, the solution was boiled for 5 min to terminate the reaction and allowed to cool to room temperature. After the filtration of the cooled mixture through a 0.45 μm filter (New Steradisc 25, Kurabo, Osaka, Japan), the resulting HA fragments were obtained by freeze-drying.

A gene carrier system (hyaluronate-grafted poly-(L-lysine) copolymer: PLL-g-HA) was prepared by coupling HA to poly-L-lysine (PLL) in a 1:1 weight [8]. The HA fragments (number-average molecular weight: 1.5×10^4), obtained as above, were conjugated to PLL (number-average molecular weight: 4.6×10^4) by reductive amination using $NaBH_3CN$ as a reductant. $NaBH_3CN$ was added to the mixture of HA (100–300 mg) and PLL (60–120 mg) in 15 ml of sodium borate buffer (0.1 M, pH 8.5) containing 0.4–1 M NaCl at a final concentration of 0.3 M. The reaction mixture was incubated at 40°C for 3 days while it was stirred constantly, followed by dialysis against a 0.5 M NaCl aqueous solution using a Spectra/Por 7 membrane (molecular weight cutoff of 25 000) to remove unreacted HA. Then, the solution was desalted by dialysis against distilled water, and the resulting copolymer was obtained by freeze-drying.

Gel permeation chromatography (GPC) was carried out using a Jasco 880-PU pumping system at the flow rate of 1.0 ml/min at 25°C with Ultrahydrogel 250 columns (Japan Waters, Tokyo, Japan). The GPC profile of the reaction mixture and the resulting copolymer showed that, after the reaction, the peak area of PLL increased while that of HA decreased. The molecular weight determined by the light scattering detector increased after the reaction. The incubation of HA or PLL alone with $NaBH_3CN$ under the same conditions caused no change in GPC profile, suggesting that neither degradation nor intermolecular side reaction occurred. When HA was incubated with PLL in the presence of 10-fold molar excess of lactose or sucrose with respect to the reducing end of HA, not sucrose, the nonreducing sugar, but lactose, the reducing one, inhibited HA consumption. Therefore it can be concluded that the HA was conjugated with PLL at its reducing end through the reductive amination reaction. The GPC profile also shows that the comb-type copolymer was isolated from unreacted HA by dialysis.

The ^1H NMR spectra of the comb-type copolymer showed the characteristic signals of both PLL and HA moieties: PLL, δ1.4–1.8 (β, γ, δ-CH_2), 3.0 (ε-CH_2), 4.3 (α-CH); HA, δ2.0 (Nac-CH_3), 3.3–3.9 (H-2,3,4,5,6), 4.4–4.6 (H-1). From the signal ratio of methyl protons (2.0 ppm) of the *N*-acetyl groups of the HA-grafts to ε-methylene protons (3.0 ppm) of the PLL backbone, the content (wt% and grafting%) of HA in the copolymer was determined. The coupling efficiency was more than about 70%. It was determined that one HA chain was grafted onto an average of one of every 150 lysine units.

Consequently, we have prepared the PLL-g-HA comb-type copolymer with a defined density and a defined length of HA side chains. The synthesis of the comb-type copolymer has been achieved by (i) the control of the HA chain length by enzymatic hydrolysis, (ii) the suppression of formation of a polyion complex between HA and PLL in high ionic strength media during coupling reaction, and (iii) the relatively high coupling efficiency of the reductive amination between the reducing end of HA and ε-amino groups of PLL.

Estimation of the PLL-g-HA/DNA Interaction

DNA was mixed with PLL-g-HA ($[\text{amino group}]_{\text{copolymer}}/[\text{phosphate group}]_{\text{DNA}} = 1$), and the ionic strength of the medium was gradually decreased from 1 to 0 M NaCl by step-down dialysis. The transparent solution obtained after dialysis was then lyophilized, dissolved in D_2H, and analyzed by ^1H NMR. The ^1H NMR spectra of the copolymer with or without DNA were analyzed and it was shown that the signals of ε-methylene protons of the PLL moiety in the PLL-g-HA broadened in the presence of DNA. This is likely caused by polyanion formation between the PLL moiety and DNA. Although the PLL signal broadened, the HA signal remained unchanged, or rather became sharper. The sharp signal of HA in the presence of DNA under low-ionic strength conditions seem to be similar to those of PLL-g-HA alone at high ionic strength, where the PLL-g-HA did not form polyion complex; i.e., HA chains were free. These results suggest that the PLL backbone selectively formed the polyanion complex with DNA even in the presence of the HA side chains.

Complex formation between PLL-g-HA and DNA was further assessed with a gel retardation assay. Increasing the proportion of the PLL-g-HA copolymer in the DNA samples affected plasmid DNA migration in an agarose gel. The titration point representing the minimum proportion of PLL-g-HA required to retard the DNA completely occurred at a 1:1 copolymer (based on PLL) to DNA charge ratio.

Based on the results, it is postulated that the PLL-g-HA/DNA complex forms the multiphase structure in which the hydrophobic PLL-g-HA complex is surrounded by the hydrated shell of free HA. The presence of free HA chains is essential for directing the complex to the target cells.

Targetability of the PLL-g-HA/DNA Complex to SECs

To determine whether the PLL-g-HA/DNA complex was recognized by SEC HA receptors in vivo, Wistar rats were injected intravenously via the tail vein with PLL-g-HA complexed to ^{32}P-labeled pSV β-Gal (an expression plasmid encoding lacZ). One hour postinjection, more than 90% of the injected radioactivity remained in the liver. Administration of the PLL-g-HA complexed to a fluorescein isothiocyanate-labeled DNA revealed that the carrier-DNA complex was distributed exclusively in SECs [9]. Polymerase chain reaction (PCR) amplification detected the presence of the transgene in the cytosolic fraction 3h after transfection only from the liver of the rats injected with PLL-g-HA/pSV β-Gal, but not with pSV β-Gal alone. A large number of cells expressing β-galactosidase were detected in SECs when transfected with the PLL-g-HA/pSV β-Gal complex.

Several lines of evidence shown in this study indicate that the complex was transferred successfully to the SECs via HA receptor-mediated endocytosis [9]. (i) Intravenous administration of the PLL-g-HA/[^{32}P]pSV β-Gal resulted in accumulation of >90% of the radioactivity in the liver. The PLL-g-HA/DNA complex was localized exclusively in the sinusoidal lining cells (i.e. SECs). (ii) The transgene was restored from the cytosolic fraction of the liver and lacZ gene expression was demonstrated in the SECs following transfection with the PLL-g-HA/pSV β-Gal complex.

Stabilization of Triplex DNA

The ability to target specific sequence of DNA through oligonucleotides-based triplex formation provides a powerful tool for genetic manipulation. Triplex-forming oligonucleotides (TFOs) can be targeted for its sequence-specific binding to the major groove of homopurine–homopyrimidine stretch in target genes. Triplex-forming oligonucleotides can inhibit or enhance DNA transcription, generate site-specific mutations, cleave DNA, and induce homologous recombination. However, instability of triplex DNA under physiological conditions, due to the electrostatic repulsion between a TFO and the duplex as well as pH-dependence and K^+ sensitivity, limits in vivo application of the triplex strategy. We previously reported that comb-type copolymers composed of PLL and hydrophilic polysaccharides significantly increased thermal stability of triplex structure.

In this study, we used PLL-g-HA, a comb-type copolymer having HA as a side chain, and evaluated its ability to stabilize triplexes. Triplex formation was analyzed by electrophoretic mobility shift assay (EMSA) using a 30-mer target duplex from rat α1 (I) collagen gene promoter (T-1/T-2 duplex, T-1: 5′-CCTTTCCCTTCCTTTCCCTCCTCCCCCCTC-3′; T-2: 3′-GGAAAGGGAAGGAAAGGGAGGAGGGGGGAG-5′); Pu-20 (5′-GGAAAGGGAAGGAAAGGGAGGAGGG-3′) and Py-20 (5′-CCCTTCCTTTCCCTCCTCCC-3′) as TFOs. PLL-g-HA diminished remarkably K^+ inhibition of purine motif triplex formation with Pu-20 as well as pH-dependence of pyrimidine motif triplex formation with Py-20 [9]. Moreover, thermal denaturation (UV-*Tm*) and circular dichroism experiments revealed that PLL-g-HA significantly increased the thermal stability of triplex structure. Thus, the PLL-g-HA/DNA carrier system stabilizes triplex DNA.

Inhibition of TNFα-Induced ICAM-1 Expression in SECs Using a Double-Strand Decoy DNA Oligomer for NF-κB

Altered gene expression of SECs is associated with impaired immune response. We determined if the decoy technique could effectively suppress tumor necrosis factor α (TNFα)-induced intercellular adhesion molecule 1 (ICAM-1) expression in SECs. A nuclear factor (NF)-κB decoy (NF-κB31: 5′-TGGGGACTTTCCAGTTTCTGGAAAGTCCCCA-3′), which contains a consensus sequence for NF-κB, was synthesized [10]. NF-κB31 was complexed to PLL-g-HA. Controls were oligodeoxynucleotides (ODNs) with reversed and scrambled sequences. The PLL-g-HA/NF-κB31 complex was added to the culture media of LSE cells, a human SEC-derived cell line, and cultured for 3 h. Then, cells were stimulated for 1–4 h with TNFα (5 ng/ml). Tumor necrosis factor α caused a dramatic translocation of NF-κB protein into the nuclei of LSE cells followed by marked upregulation of ICAM-1 on the cell surface. PLL-g-HA/NF-κB31, but not control ODNs, inhibited the intranuclear localization of NF-κB induced by TNFα, with almost complete inhibition at 2.5 μg/ml. NF-κB31 attenuated the increase in ICAM-1 mRNA as well as protein levels in LSE cells. The decoy technique in combination with PLL-g-HA may provide a novel strategy for manipulation of SEC functions.

Conclusion

The PLL-g-HA/DNA carrier system permits targeted delivery of exogenous genes selectively to the liver SECs. The gene carrier system in combination with the decoy or triplex DNA approaches may lead to a new and more effective strategy for manipulation of SEC functions.

Acknowledgments. This work was supported by the Grants-in-Aid for the Development of Scientific Research (B 07557045 and C2 15590690) from the Ministry of Education and Science, Japan, and by grants from the Uehara Memorial Foundation and Kanae Medical Foundation.

References

1. Mantovani A, Bussolino F, Dejana E (1992) Cytokine regulation of endothelial cell function. FASEB J 6:2591–2599
2. Pober JS, Cotran RS (1990) The role of endothelial cells in inflammation. Transplantation 50:537–544
3. Takei Y, Marzi I, Gao W, et al (1991) Leukocyte adhesion and cell death following orthotopic liver transplantation in the rat. Transplantation 51:959–965
4. Bach FH, Robson SC, Winkler H, et al (1995) Barriers to xenotransplantation. Nat Med 1:869–873
5. Bach FH, Robson SC, Ferran C, et al (1994) Endothelial cell activation and thromboregulation during xenograft rejection. Immunol Rev 141:5–30
6. Politz O, Gratchev A, McCourt PA, et al (2002) Stabilin-1 and -2 constitute a novel family of fasciclin-like hyaluronan receptor homologues. Biochem J 362:155–164
7. Zhou B, Weigel JA, Saxena A, et al (2002) Molecular cloning and functional expression of the rat 175-kDa hyaluronan receptor for endocytosis. Mol Biol Cell 13:2853–2868
8. Asayama S, Nogawa M, Takei Y, et al (1998) Synthesis of novel polyampholyte comb-type copolymers consisting of a poly(L-lysine) backbone and hyaluronic acid side chains for a DNA carrier. Bioconjug Chem 9:476–481
9. Takei Y, Maruyama A, Ferdous A, et al (2004) Targeted gene delivery to sinusoidal endothelial cells: DNA nanoassociate bearing hyaluronan-glycocalyx. FASEB J 2004; 18:699–701 (10.1096/fj.03-0494fje)
10. Shibuya T, Takei Y, Hirose M, et al (2002) A double strand DNA oligomer having a consensus sequence for NF-κB (decoy) to inhibit TNFα-induced ICAM-1 expression in sinusoidal endothelial cells. Biochem Biophys Res Commun 298:10–16

Basic Aspects of the Hepatic Microvascular System

Robert S. McCuskey

The hepatic microvascular system comprises all blood and lymphatic vessels immediately involved in the delivery and removal of fluids to and from the hepatic parenchyma. The majority of blood enters the sinusoids (exchange vessels) from portal venules, the remainder from hepatic arterioles. Blood leaves the sinusoids by flowing into central venules. Arterial blood enters some of the sinusoids through branches of the hepatic arterioles. Occasional direct connections also have been observed with the portal venules and post-sinusoidal vessels. Because all of these vessels are independently contractile, the sinusoids receive a varying mixture of portal venous and arterial blood.

The sinusoid network exhibits structural and dimensional heterogeneity. Near portal venules, sinusoids are arranged in interconnecting polygonal networks; further away from the portal venules, the sinusoids become organized as parallel vessels that terminate in the central venules. In addition, the diameter of periportal sinusoids is less than those in the centrilobular region. The ultrastructural fenestration of the sinusoids also exhibits regional heterogeneity.

The functional unit is thought to be a conical "hepatic microvascular subunit" of the classic lobule consisting of a group of sinusoids supplied by a single inlet venule and its associated termination of a branch of the hepatic arteriole and cholangiole.

There are several potential morphological sites for regulating blood flow through the sinusoids. These include the various segments of the afferent portal venules and hepatic arterioles, the sinusoids themselves, and the central and hepatic venules. The principal site of regulation of blood flow

Department of Cell Biology and Anatomy, College of Medicine, University of Arizona Tucson, AZ, USA through the sinusoids, however, is thought to reside in the sinusoid itself, where the major blood pressure drop occurs in the liver. Sinusoidal lining cells produce, and are responsive to, a variety of substances. Sinusoidal endothelial cells may act like sphincters by swelling or contracting, thereby narrowing the sinusoidal lumen and limiting blood flow. More recently, attention has focused on the perisinusoidal stellate cell as the cell responsible for controlling sinusoidal diameter and blood flow.

Activation of Hepatic Stellate Cells: Anti-Adipogenic Differentiation?

Hidekazu Tsukamoto

Summary. Hepatic stellate cells (HSC) are vitamin-A storing, liver pericytes that become transdifferentiated into myofibroblastic cells to play a central role in liver fibrogenesis. Elucidation of molecular mechanisms underlying this transdifferentiation represents the most pivotal prerequisite for development of efficacious therapeutic modalities for cirrhosis and portal hypertension. To this end, our laboratory proposed a notion that HSC differentiation is analogous to adipocyte differentiation. In support of the notion, the depletion of peroxisome proliferator-activated receptor (PPAR) γ, a master transcription factor for adipocyte differentiation, is closely associated with myofibroblastic transdifferentiation of HSC and conversely the treatment of the cells with ligands for PPARγ or forced expression of this nuclear factor reverses the phenotype of "activated" HSC to that of quiescent HSC. Further, the expression of other known adipogenic transcription factors are most robust in quiescent HSC and are drastically diminished in activated HSC. This proposal raises an intriguing possibility that pro-adipogenic manipulation targeted to HSC may serve as a unique therapeutic intervention for liver fibrosis and hepatic microcirculation disturbances.

Key words. Peroxisome proliferator-activated receptor γ, Preadipocyte, Adipogenesis, Liver fibrosis

---

Research Center for Alcoholic Liver and Pancreatic Diseases, Keck School of Medicine of the University of Southern California, 1333 San Pablo Street, MMR412, Los Angeles, CA 90033, USA

Molecular Investigation of Hepatic Stellate Cell Activation

Hepatic stellate cells (HSC) are pericytes located in the subendothelial space along the hepatic sinusoid, the terminal capillary in the liver. They extend dendritic cytoplasmic extensions so as to encircle the wall of the sinusoid to fulfill the role of pericytes. In addition, HSC are the principal sites of vitamin A storage and help maintain the differentiated phenotype of hepatocytes via communication through gap junctions, release of soluble factors, and production of normal perisinusoidal matrix constituents. Upon injury to the liver, HSC undergo cellular transdifferentiation characterized by the loss of vitamin A, increased proliferation, increased expression of chemokines, cytokines, and adhesion molecules, as well as induction of matrix metalloproteinases (MMPs), matrix proteins, and tissue inhibitors of metalloproteinases (TIMPs) [1,2]. An additional and important feature of HSC "activation" is acquisition of expression for α-smooth muscle actin that confers upon the cells a contractile property and a myofibroblastic phenotype. This cellular activation, if uncontrolled, leads to a fibroproliferative disease of the liver and portal hypertension resulting from disturbed microcirculation. No curative therapy is currently available for cirrhosis except for liver transplantation. Besides alcoholic liver disease, highly prevalent viral hepatitis and nonalcoholic steatohepatitis (NASH) associated with obesity and type II diabetes, contribute to the incidence and complications of cirrhosis, one of the leading causes of medical morbidity and mortality in our society. To this end, much research effort has been rendered to elucidate key steps in transcriptional regulation that is responsible for HSC transdifferentiation. Identification of such molecular targets is crucial for development of new medical treatments for the disease. Such examples include Kruppel-like factor (KLF)-6, a differentially induced zinc finger protein in activated HSC, that binds to the GC box sites of several key fibrogenic genes such as transforming growth factor (TGF) β1, TGFβ receptor type I and II [3], urokinase-type plasminogen activator [4], and α1(I) procollagen [3]. Induction of α-smooth muscle actin is transcriptionally mediated by c-Myb binding to an E-box element in its promoter [5]. Activation of NF-κB may support the proliferative status of activated HSC [6] while it also mediates their participation in inflammatory and immunoregulatory responses via induction of chemokines, adhesion molecules, and immunoregulatory molecules [7]. Increased AP-1 activity is essential for induction of MMPs, TIMP-1, and interleukin (IL)-6 gene transcription in activated HSC in which JunD is shown to play a central role [8].

Analogy Between Adipogenic Differentiation and HSC Differentiation

A difficulty in understanding HSC differentiation is highlighted by diverse cellular phenotype markers that HSC express. Besides the myofibroblastic phenotype, HSC also express MyoD, the myogenic transcription factor specific for skeletal muscle [9]. Neural markers such as glial fibrillary acidic protein (GFAP) [10], neural cell adhesion molecule (N-CAM) [11], nestin [11], and synaptophysin [12] are also expressed in HSC. Activated HSC express leptin [13], an adipocyte specific gene, raising an interesting possibility that HSC may also share the adipocyte phenotype. The latter notion is also supported by the expression by activated HSC of other "adipokines," the proteins expressed by adipocytes. These include osteonectin (SPARC) [14] and plasminogen activator inhibitor-1 (PAI-1) [15]. In a more simplistic context, the quiescent HSC stores retinyl esters but are also laden with lipids such as triglycerides, cholesterol, and phospholipids [16] much like adipocytes. The mediators known to activate HSC such as TGFβ platelet-derived growth factor (PDGF), TGFα, tumor necrosis factor (TNF) α, and leptin, also inhibit adipocyte differentiation and promote pre-adipocyte differentiation (Fig. 1). In line with this notion, Miyahara et al. demonstrated in our laboratory that

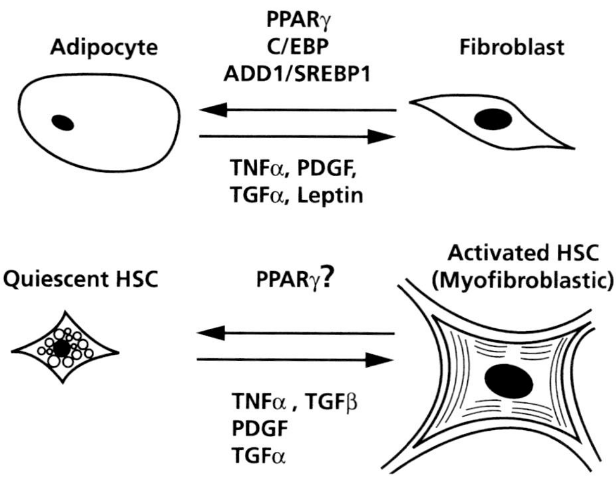

Fig. 1. A schematic diagram depicting the analogy between adipocyte differentiation and hepatic stellate cell (*HSC*) differentiation. *PPARγ*, peroxisome proliferator-activated receptor γ; *C/EBP*, CAAT-enhancer-binding protein; *ADD1*, adipocyte determination and differentiation factor 1; *SREBP1*, sterol regulatory element binding protein 1; *TNFα*, tumor necrosis factor α; *TGFα*, transforming growth factor α; *TGFβ*, transforming growth factor β; *PDGF*, platelet-derived growth factor

Table 1. Similarity between adipocyte differentiation and HSC differentiation

Parameter	Adipocyte	HSC
Triglyceride content	×	×
Fibroblastic transdifferentiation	×	×
Expression of adipokines		
Leptin	×	×
Osteonectin	×	×
PAI-1	×	×
Expression of adipogenic transcription factors		
PPARγ	×	×
C/EBP	×	×
Mediators of transdifferentiation		
TGFα	×	×
PDGF	×	×
TGFβ	×	×
TNFα	×	×
Leptin	×	×

HSC, hepatic stellate cells; PPAR, peroxisome proliferator-activated receptor; C/EBP, CAAT-enhancer-binding protein; PAI, plasminogen activator inhibitor; TGF, transforming growth factor; PDGF, platelet-derived growth factor; TNF, tumor necrosis factor

quiescent HSC express peroxisome proliferator-activated receptor γ (PPARγ), a key transcription factor for adipocyte differentiation [17], and this nuclear receptor is depleted in activated HSC in vitro [17,18] and in vivo [17]. Moreover, the treatment of culture-activated HSC with natural or synthetic ligands for PPARγ, inhibits the HSC activation markers such as cell proliferation [18], expression of collagen, TGFβ, α-smooth muscle actin, MCP-1 genes, and chemotaxis [17,19]. These studies have eventually led to a demonstration of antifibrotic efficacy of PPARγ ligands in animal models of liver fibrosis [18]. These findings have a direct relevance to the therapeutic potential of PPARγ ligands but also shed fundamental insights into the understanding of the regulatory mechanisms underlying HSC transdifferentiation, i.e., the maintenance of the quiescent state of HSC requiring PPARγ, as adipocyte differentiation does. The similarity between adipocyte differentiation and HSC differentiation is summarized in Table 1.

Receptor-Independent Effects of PPARγ Ligands

Antifibrotic effects of PPARγ ligands are also shown for fibroproliferative diseases of other organs including artherosclerosis [20], glomerular sclerosis [21], and arthritis [22]. They also exert anti-inflammatory effects on macrophages by suppressing their expression of cytokines [23] while promoting CD36 expression and differentiation of monocytes/macrophages [24]. However, these biological effects of the ligands may be mediated via PPARγ-independent mechanisms. The best example that demonstrates this receptor independency is anti-inflammatory effects of the synthetic ligands on macrophages derived from embryonic stem cells of PPARγ null mice [25]. In fact, we now know that the PPARγ ligand 15-deoxy-prostaglandin J_2 (15dPGJ2) inhibits NF-κB activation by directly inhibiting IκB kinase in a PPARγ-independent manner [26]. Troglitazone, the first generation of thiazolidinedione (TZD) derivatives, also selectively induces early growth response-1 gene independently of PPARγ [27]. MAP kinases are activated by 15dPGJ2 and ciglitazone, another TZD in astrocytes and preadipocytes via the mechanisms that are independent of PPARγ but involving reactive oxygen species [28]. Thus, when one considers the effects of PPARγ ligands on HSC and liver fibrosis, it is critical to ask whether they are mediated via the nuclear receptor. To directly test the role of PPARγ in regulation of HSC differentiation, one can perform a gain-of-function experiment. This was recently implemented by Hazra et al. in our laboratory [29]. As discussed above, culture-activated primary rat HSC are depleted of PPARγ expression. To restore the expression of the receptor, we have employed an adenoviral vector. This manipulation caused a phenotypic switch from activated HSC to the quiescent phenotype with a reversal of the morphological characteristics and suppressed fibrogenic markers such as the expression of collagen, TGFβ and α-smooth muscle actin. More importantly, the cells regained the ability to store vitamin A [29]. Furthermore, this study identified a physical interaction between PPARγ and JunD as the potential molecular mechanism underlying the reversal effects of PPARγ [29].

Pro-Adipogenic Mode of HSC Differentiation and Antifibrotic Modalities

The notion that HSC differentiation is governed by the regulatory mechanisms known for adipocyte differentiation is further supported by our recent demonstration of other adipogenic transcription factors that are also expressed at high levels in quiescent HSC and are reduced upon culture activation [30]. More importantly, cultured-activated HSC show a reversal to the quiescent phenotype upon the treatment with the adipocyte differentiation cocktail that contains methylisobutylxanthine, dexamethasone, and insulin. Further, this effect ensues in parallel with the restoration of the expression for the adipogenic transcription factors [30]. These results reinforce the concept that pro-adipogenic regulation of HSC may indeed serve as an effective anti-fibrotic modality via promotion of HSC differentiation. However, an intervention that induces PPARγ nonselectively in the liver may be problematic since upregulation of PPARγ in hepatocytes with suppressed PPARα activity leads to induction of fatty liver [31], and suppressed PPARα activity is a common observation in chronic liver disease. Thus, the PPARγ modality has to be directed specifically to HSC to have an efficacious effect with minimal side effects on hepatocytes. Alternatively, one may consider modalities to sensitize insulin signaling. Adiponectin may fall into this category of new potential modalities for liver fibrosis [32].

Acknowledgments. The work described in this article was supported by National Institutes of Health Grants R37AA06603, P50AA11999 (USC-UCLA Research Center for Alcoholic Liver and Pancreatic Diseases), P30DK48522 (USC Research Center for Liver Diseases), and R24AA12885 (Non-Parenchymal Liver Cell Core) and by the Medical Research Service of the Department of Veterans Affairs.

References

1. Hautekeete ML, Geerts A (1997) The hepatic stellate (Ito) cell: its role in human liver disease. Virchows Arch 430:195–207
2. Friedman SL (1996) Hepatic stellate cells. Prog Liver Dis 14:101–130
3. KimY, RatziuV, Choi SG, et al (1998) Transcriptional activation of transforming growth factor beta1 and its receptors by the Kruppel-like factor Zf9/core promoter-binding protein and Sp1. Potential mechanisms for autocrine fibrogenesis in response to injury. J Biol Chem 273:33750–33758
4. Kojima S, Hayashi S, Shimokado K, et al (2000) Transcriptional activation of urokinase by the Kruppel-like factor Zf9/COPEB activates latent TGF-beta1 in vascular endothelial cells. Blood 95:1309–1316
5. Buck M, Kim DJ, Houglum K, et al (2000) c-Myb modulates transcription of the alpha-smooth muscle actin gene in activated hepatic stellate cells. Am J Physiol Gastrointest Liver Physiol 278:G321–G328
6. Lang A, Schoonhoven R, Tuvia S, et al (2000) Nuclear factor kappaB in proliferation, activation, and apoptosis in rat hepatic stellate cells. J Hepatol 33:49–58
7. Hellerbrand C, Jobin C, Licato LL, et al (1998) Cytokines induce NF-kappaB in activated but not in quiescent rat hepatic stellate cells. Am J Physiol 275:G269–G278
8. Smart DE, Vincent KJ, Arthur MJ, et al (2001) JunD regulates transcription of the tissue inhibitor of metalloproteinases-1 and interleukin-6 genes in activated hepatic stellate cells. J Biol Chem 276:24414–24421
9. Vincent KJ, Jones E, Arthur MJ, et al (2001) Regulation of E-box DNA binding during in vivo and in vitro activation of rat and human hepatic stellate cells. Gut 49:713–719
10. Neubauer K, Knittel T, Aurisch S, et al (1996) Glial fibrillary acidic protein—a cell type specific marker for Ito cells in vivo and in vitro. J Hepatol 24:719–730
11. Niki T, Pekny M, Hellemans K, et al (1999) Class VI intermediate filament protein nestin is induced during activation of rat hepatic stellate cells. Hepatology 29:520–527

12. Cassiman D, van Pelt J, De Vos R, et al (1999) Synaptophysin: a novel marker for human and rat hepatic stellate cells. Am J Pathol 155:831–1839
13. Potter JJ, Womack L, Mezey E, et al (1998) Transdifferentiation of rat hepatic stellate cells results in leptin expression. Biochem Biophys Res Commun 244:78–182
14. Nakatani K, Seki S, Kawada N, et al (2002) Expression of SPARC by activated hepatic stellate cells and its correlation with the stages of fibrogenesis in human chronic hepatitis Virchows Arch 441:466–474
15. Knittel T, Fellmer P, Ramadori G (1996) Gene expression and regulation of plasminogen activator inhibitor type I in hepatic stellate cells of rat liver. Gastroenterology 111:45–754
16. Yamada M, Blaner WS, Soprano DR, et al (1987) Biochemical characteristics of isolated rat liver stellate cells. Hepatology 7:1224–1229
17. Miyahara T, Schrum L, Rippe R, et al (2000) Peroxisome proliferator-activated receptors and hepatic stellate cell activation. J Biol Chem 275:35715–35722
18. Galli A, Crabb DW, Ceni E, et al (2002) Antidiabetic thiazolidinediones inhibit collagen synthesis and hepatic stellate cell activation in vivo and in vitro. Gastroenterology 122:1924–1940
19. Marra F, Efsen E, Romanelli RG, et al (2000) Ligands of peroxisome proliferator-activated receptor gamma modulate profibrogenic and proinflammatory actions in hepatic stellate cells. Gastroenterology 119:466–478
20. Chen Z, Ishibashi S, Perrey S, et al (2001) Troglitazone inhibits atherosclerosis in apolipoprotein E-knockout mice: pleiotropic effects on CD36 expression and HDL. Arterioscler Thromb Vasc Biol 21:372–377
21. Buckingham RE, Al Barazanji KA, Toseland CD, et al (1998) Peroxisome proliferator-activated receptor-gamma agonist, rosiglitazone, protects against nephropathy and pancreatic islet abnormalities in Zucker fatty rats. Diabetes 47:1326–1334
22. Kawahito Y, Kondo M, Tsubouchi Y, et al (2000) 15-deoxy-delta(12,14)-PGJ(2) induces synoviocyte apoptosis and suppresses adjuvant-induced arthritis in rats. J Clin Invest 106:189–197
23. Jiang C, Ting AT, Seed B (1998) PPAR-gamma agonists inhibit production of monocyte inflammatory cytokines. Nature 391:82–86
24. Tontonoz P, Nagy L, Alvarez JG, et al (1998) PPARgamma promotes monocyte/macrophage differentiation and uptake of oxidized LDL. Cell 93:241–252
25. Chawla A. Barak Y, Nagy L, et al (2001) PPAR-gamma dependent and independent effects on macrophage-gene expression in lipid metabolism and inflammation Nat Med 7:48–52
26. Straus DS, Pascual G, Li M, et al (2000) 15-deoxy-delta 12,14-prostaglandin J2 inhibits multiple steps in the NF- kappa B signaling pathway. Proc Natl Acad Sci USA 97:4844–4849
27. Baek SJ, Wilson LC, His LC, et al (2003) Troglitazone, a peroxisome proliferator-activated receptor gamma (PPAR gamma) ligand, selectively induces the early growth response-1 gene independently of PPAR gamma. A novel mechanism for its anti-tumorigenic activity. J Biol Chem 278:5845–5853
28. Lennon AM, Ramauge M, Dessouroux A, et al (2002) MAP kinase cascades are activated in astrocytes and preadipocytes by 15-deoxy-Delta(12–14)-prostaglandin J(2) and the thiazolidinedione ciglitazone through peroxisome proliferator activator receptor gamma- independent mechanisms involving reactive oxygenated species. J Biol Chem 277:29681–29685

29. Hazra S, Xiong S, Wang J, et al (2004) Peroxisome proliferator-activated receptor gamma induces a phenotypic switch from activated to quiescent hepatic stellate cells. J Biol Chem 279:11392–11401
30. She H, Hazra S, Xiong S, et al (2003) Adipogenic regulation of hepatic stellate cell transdifferentiation. Hepatology 38:258A
31. Yu S, Matsusue K, Kashireddy P, et al (2003) Adipocyte-specific gene expression and adipogenic steatosis in the mouse liver due to peroxisome proliferator-activated receptor gamma1 (PPARgamma1) overexpression. J Biol Chem 278:498–505
32. Kamada Y, Tamura S, Kiso S, et al (2003) Enhanced carbon-tetrachloride induced liver fibrosis in mice lacking adiponectin. Gastroenterology 125:1796–1807

Microvascular Derangement in Liver Transplantation

Michael D. Menger

Microcirculatory derangements in organ transplantation, characterized by capillary perfusion failure and inflammation-associated leukocyte recruitment, are major determinants of the manifestation of graft dysfunction and destruction. Although preservation/cold storage, post-transplant reperfusion, and rejection have to be considered as individual factors contributing to injury, recent studies have indicated that ischemia/reperfusion-associated events may trigger immune-response-mediated late rejection. There is major evidence that the microcirculatory derangements induced by cold preservation and reperfusion involve oxygen radicals, complement, phospholipase A2, leukotrienes, thromboxane, platelet-activating factor and endothelin-1, as well as the activation and functioning of leukocytic and endothelial selectins, β2-integrins, and ICAM-1. This view is based on the fact that blockade or neutralization of these inflammatory mediators and adhesion molecules results in significant amelioration of microvascular graft dysfunction. Interestingly, the ischemia/reperfusion-associated apoptotic cell death additionally contributes to the manifestation of microcirculatory derangements, because inhibition of apoptosis by blockade of p53 results in amelioration of microcirculation dysfunction. The introduction of novel techniques for the study of human microcirculation, such as thermodiffusion and orthogonal polarization spectral imaging, may in future assist in improving both early diagnosis of microcirculatory derangements and in the monitoring of the appropriateness of therapy in clinical transplantation surgery.

Institute for Clinical and Experimental Surgery, University of Saarland, D-66421 Homburg/Saar, Germany

Regulation of Local Hepatic Oxygen Delivery Following Stress

Mark G. Clemens, Markus Paxian, Walid Kamoun, Jean Ashburn, M. Nicole Kresge, and Sandra Merkel

Summary. Hepatic dysfunction is a common sequela of a wide variety of stresses such as ischemia/reperfusion, sepsis, chronic alcohol consumption and even remote trauma. Hypersensitivity to the constrictor activity of endothelin (ET)-1 is a common finding in the liver microcirculation in all of these stresses. The mechanism of this increased constrictor response appears to be dependent upon an increased expression of ET_B receptors but with concomitant uncoupling of the ET_B receptor from activation of endothelial nitric oxide synthase (eNOS). This uncoupling is at least in part the result of over-expression of caveolin-1 which binds to eNOS thus preventing activation by calmodulin. Increased expression of caveolin is associated with decreased basal eNOS activity and profound inhibition of endothelin-stimulated eNOS activity. Finally, uncoupling of ET_B binding from eNOS activation leads to local tissue hypoxia and potentiation of injury. We propose that this mechanism is an important contributor to hepatic dysfunction following stress.

Key words. Endothelin, Caveolin-1, Microcirculation, Intravital microscopy

Introduction

A wide variety of stresses involving the liver, including inflammation such as following sepsis or oxidative stress associated with ischemia and reperfusion or hemorrhagic shock, lead to deficits in oxygen utilization by the liver. These deficits are caused by a combination of failure of the liver cells to effectively utilize available oxygen and changes in vascular regulation, resulting in areas

---

Department of Biology, University of North Carolina at Charlotte, 9201 University City Boulevard, Charlotte, NC 28223, USA of compromised perfusion. Although there is controversy regarding which of these mechanisms are the primary mechanism leading to hepatic injury in these conditions, work from our laboratory as well as others have demonstrated that the ability to effectively restore microvascular perfusion is a critical determinant of the recovery of the liver [1]. Severe injury such as occurs after prolonged ischemia and reperfusion results in structural damage to the vasculature and rapid failure of the microcirculation. In addition, inflammatory stresses result in neutrophil accumulation in the liver, which can also result in parenchymal cell injury as well as vascular failure. However, deficits in oxygen extraction also occur in the absence of overt injury to the vasculature. In these cases data suggest that altered reactivity of the blood vessels to vasoconstrictors and vasodilators contributes to deficits in oxygen delivery [2,3].

Regulation of the liver microcirculation is complex. The liver has a dual circulation. While the hepatic artery circulation is regulated similar to most vascular beds, the volumetric flow of the portal circulation is regulated largely by the upstream splanchnic viscera. These two circulations join to perfuse the sinusoids. Although the sinusoids, like capillaries in other vascular beds, do not contain vascular smooth muscle cells, it is now commonly accepted that they are subject to active constriction [4,5]. This constriction is mediated by the contraction of a specialized pericyte, the hepatic stellate cell (HSC) [5,6]. Under normal conditions, the HSC contracts in response to specific peptide mediators such as endothelin, but not α-adrenergic agonists such as phenylephrine [5]. They also relax in response to nitric oxide (NO) and carbon monoxide (CO). In response to long-term stresses leading to fibrosis, the HSCs become activated and take on a myofibroblast-like phenotype. In this activated state they become highly contractile [7].

Even following relatively short-term stresses that do not result in stellate cell transformation to myofibroblasts, the microvasculature of the liver becomes hyper-responsive to the constrictor effect of endothelin (ET)-1 [8,9]. Endothelin expression is upregulated following endotoxemia, hypoxia, or ischemia [10]. In addition, the response to exogenous endothelin is increased following endotoxemia, ischemia reperfusion, blunt trauma, and chronic alcohol consumption [11–13]. This ubiquitous occurrence of hypersensitivity to endothelin associated with such conditions suggests that it is a common pathway to vascular dysregulation following most stresses that result in liver injury. The mechanisms for this increased sensitivity to endothelin and the functional significance has, however, been unclear. The focus of this chapter will be to present the mechanisms related to altered hepatic vascular control following inflammatory or oxidative stress and the functional importance of this altered regulation related to oxygen delivery to the liver tissue.

Endothelin Receptors

Endothelin acts through two major receptor subtypes, ET_A and ET_B. In the liver, the ET_A receptors are distributed primarily to the hepatocytes, stellate cells and vascular smooth muscle cells, while ET_B receptors are ubiquitously distributed over all cell types. The two receptor subtypes on different cells allow endothelin to exert both constrictor and dilator influences. Classically, ET_A receptors are considered to mediate constriction because of their distribution on vascular smooth muscle cells. ET_B receptors, conversely, mediate dilation by being coupled to activation of endothelial nitric oxide synthase (eNOS) (Fig. 1). Based on these functional characteristics, we originally hypothesized that an upregulation of ET_A receptors could account for the hyperconstrictive response to endothelin. However, our previous studies have shown that although both endothelin itself and overall endothelin receptors are upregulated in response to stresses such as endotoxin or ischemia/reperfusion, the predominant receptor subtype that is responsible for the increase in total receptors is the ET_B [6,14] receptor. This raised the question of how could an increase in the receptors that mediate dilation result in an enhancement of the constrictor response. Two major observations suggested a mechanism. First, although our original report did not find any sinusoidal

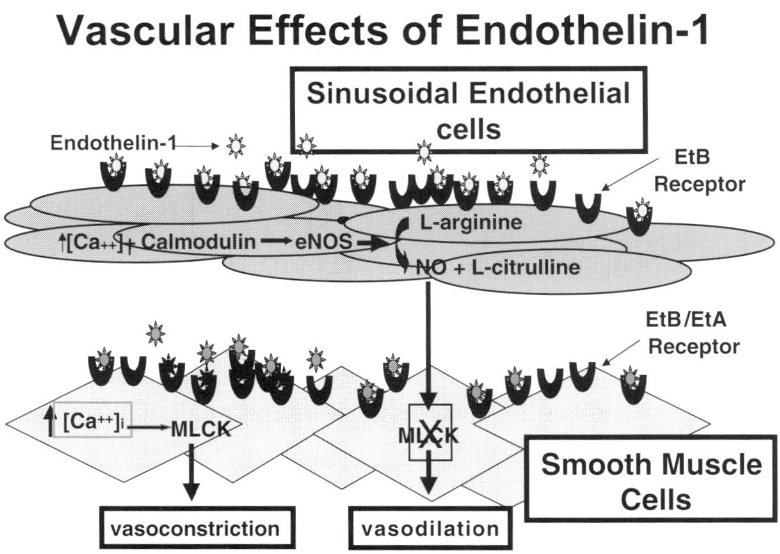

FIG. 1. Schematic diagram of the interaction between ET_A and ET_B receptors on endothelial cells and smooth muscle cells. In the liver, receptor distribution on hepatic stellate cells is similar to that of vascular smooth muscle cells. *MLCK*, myosin light chain kinase

constrictor response to ET_B receptor activation, ET_B agonists did have potent constrictor effects in the liver with the site of action being extrasinusoidal [5]. Second, it was possible to unmask the sinusoid constrictor response to an ET_B agonist by simultaneously inhibiting eNOS [6]. This observation demonstrated that there is a functional link between ET_B receptors and NOS activation in the liver, and suggested that interruption of this link may give rise to an increase in the functional constrictor response via inhibition of the compensatory dilation.

Caveolin-1 and Activation of eNOS

A possible mechanism for uncoupling of ET_B receptor binding and the activation of eNOS can be found in the putative role of caveolin-1 in eNOS regulation. Recent studies from several groups have provided evidence that increased constrictor tone in the hepatic portal circulation may be the result of decreased ability to activate eNOS. Rockey's group showed, somewhat paradoxically, that eNOS protein expression was normal but enzymatic activity was decreased in a cirrhosis model. A probable mechanism for this discrepancy was suggested by the work of Shah's group which showed that the membrane scaffolding protein caveolin-1 was upregulated in cirrhosis and that eNOS was found to be associated with the caveolin [15,16].

Caveolin is a protein associated with membrane subdomains called caveolae. Caveolar domains are particularly rich in cholesterol and sphingolipids, and are a site of clustering of receptors and signal transduction proteins. Both ET_B receptors and eNOS have been shown to associate with caveolae. The role of caveolae in regulating eNOS is particularly relevant. It is thought that binding to caveolin is necessary to locate the eNOS molecule to this subdomain where it is available for activation. On the other hand, binding to caveolin appears to inhibit binding to activated calmodulin and, thus, inhibits activation of eNOS.

These observations suggested that an increased expression of caveolin-1 in the liver might serve to uncouple ET_B receptor binding from activation of eNOS. We have examined this possibility is several models, including endotoxemia and remote trauma. Injection of 1 mg/kg *Escherichia coli* lipopolysaccharide (LPS) results in an approximately twofold increase in caveolin protein in the liver. Immunohistochemistry studies indicated that the upregulation is largely associated with vascular tissue. To test whether the sinusoidal endothelial cells might be the site of the upregulation of caveolin-1, we treated isolated sinusoidal endothelial cells with LPS and measured induction of caveolin-1 protein by Western blot. Within 6h, caveolin-1 levels were increased greater than twofold compared to vehicle controls. These results suggested that bacterial LPS upregulates caveolin-1 in sinusoidal endothelial cells in vivo. Moreover, this response was found not to be affected by the omission of serum from the medium, suggesting that the mechanism is independent of CD14 activation. Since we also found that sinusoidal endothelial cells express toll like receptor 4 (TLR4), we have hypothesized that this pathway is important for the induction of caveolin in the liver.

We have also demonstrated that caveolin association with eNOS increases following stresses. This finding is consistent with the hypothesis that increased caveolin expression serves to inhibit the activation of eNOS by calmodulin. In support of this hypothesis, we have also found that eNOS activity (but not protein expression) is decreased following remote trauma in the whole liver, and endothelin-stimulated eNOS activity is decreased following treatment of isolated sinusoidal endothelial cells with LPS. Taken together, these results support the hypothesis that increased caveolin expression inhibits activation of eNOS by ET_B receptor activation, leading to a hyperconstriction response (see Fig. 2). This then raised the question of whether uncoupling of ET_B receptor binding from eNOS activation leads to functional alterations in tissue oxygen delivery.

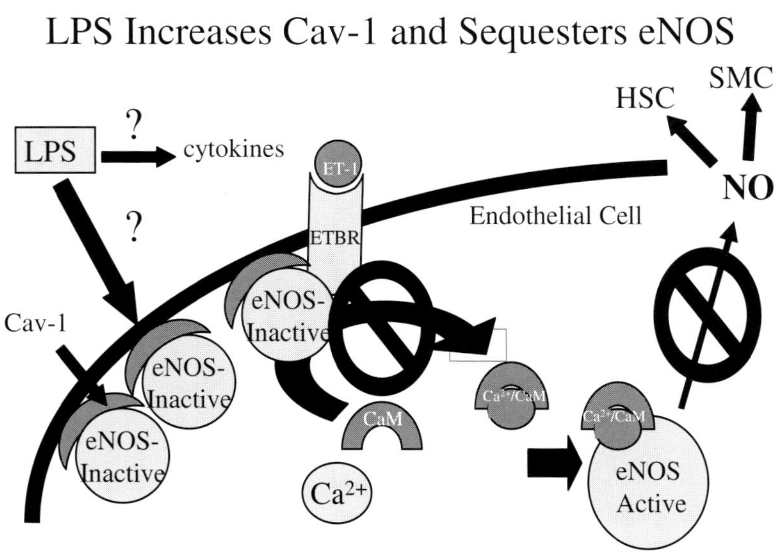

FIG. 2. Proposed schema for regulation of response to endothelin B (ET_B) receptor binding following endotoxemia. *LPS*, lipopolysaccharide; *Cav-1*, caveolin-1; *ETBR*, endothelin B receptor; *eNOS*, endothelial nitric oxide synthase; *SMC*, smooth muscle cells; *HSC*, hepatic stellate cells

Effect of Endothelins on Oxygen Delivery

To examine the effect of uncoupling ET_B receptors from NOS activation on tissue oxygenation, we used an intravital microscopy technique to examine tissue oxygenation and redox potential. Figure 3 shows the redox potential response of the liver (as indicated by nicotinamide adenine dinucleotide (NADH) fluorescence) to ET-1 which stimulates both ET_A and ET_B receptors, versus IRL1620, which is a specific ET_B agonist. The results showed that IRL 1620 exerted modest effects on redox potential compared to ET-1. This is consistent with the ET_A activation by ET-1 causing sinusoidal constriction, while the response to IRL 1620 is balanced by NO-mediated dilation. In contrast, when NOS activity was inhibited by treatment with L-nitroarginine, IRL 1620 provoked a significantly enhanced response, similar to that produced by ET-1. Similar responses were observed regarding tissue PO_2 (Fig. 4). These results are also consistent with our observation that even though both ET-1 and IRL 1620 stimulate increased oxygen demand in isolated hepatocytes, ET-1, but

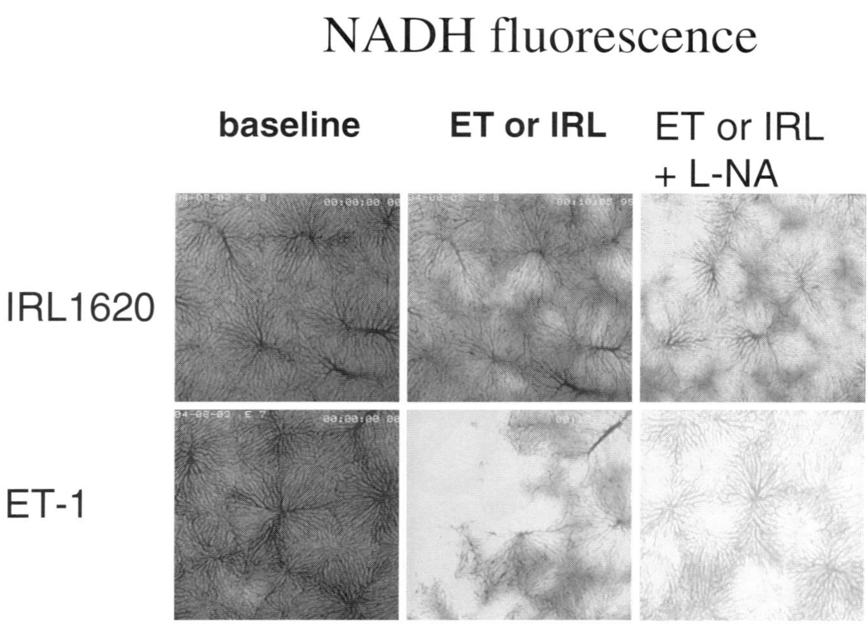

FIG. 3. Nicotinamide adenine dinucleotide (*NADH*) fluorescence in response to ET-1 or IRL 1620 without or with simultaneous treatment with L-nitroarginine (*L-NA*). L-Nitroarginine treatment alone did not cause significant change in NADH fluorescence. Rat livers were observed by intravital fluorescence microscopy in vivo. Endothelin-1 or IRL 1620 were infused via the portal vein

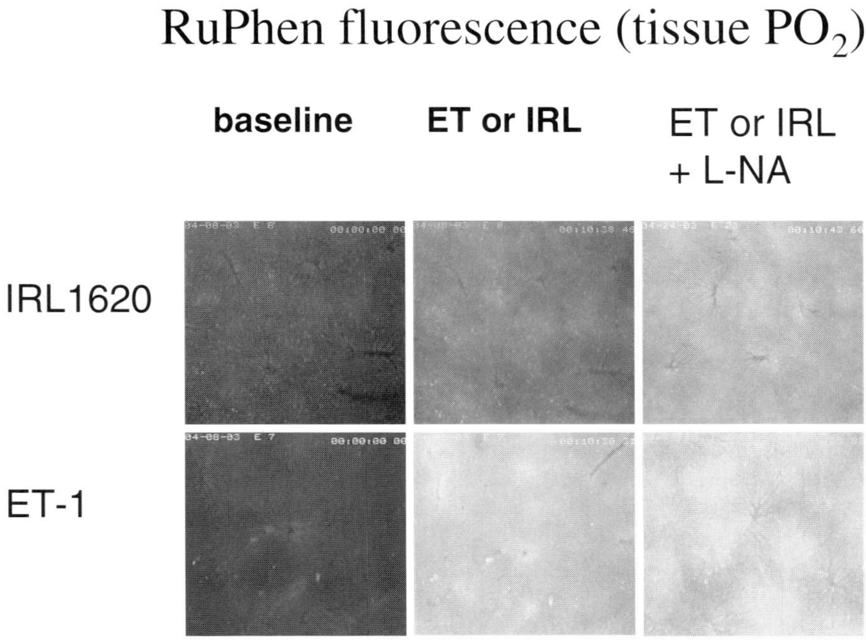

Fig. 4. Tissue PO_2 in response to ET-1 or IRL 1620. The experiment was the same as in Fig. 3 except that tris(1,10-phenanthroline) ruthenium(II) chloride hydrate (RuPhen) was infused intravenously as an indicator of tissue PO_2. Increased fluorescence indicates hypoxia

not IRL 1620, actually produces a decrease in oxygen consumption in isolated perfused liver, suggesting inadequate oxygen delivery to the tissue. This notion is further substantiated by the observation that inhibition of eNOS combined with either IRL 1620 or ET-1 potentiates cell injury, as indicated by increased enzyme release.

Conclusion

Hypersensitivity to endothelin-1 is a common finding in the liver microcirculation following a wide variety of stresses. The mechanism of this increased constrictor response appears to be dependent upon an increased expression of ET_B receptors but with concomitant uncoupling of the ET_B receptor from activation of eNOS. This uncoupling is at least in part the result of overexpression of caveolin-1, which binds to eNOS thus preventing activation by calmodulin. Increased expression of caveolin is associated with decreased basal eNOS activity and profound inhibition of endothelin-stimulated eNOS activity. Finally, uncoupling of ET_B binding from eNOS activation leads to local tissue hypoxia and potentiation of injury. We propose that this mechanism is an important contributor to hepatic dysfunction following stress.

References

1. Chun K, Zhang J, Biewer J, et al (1994) Microcirculatory failure determines lethal hepatocyte injury in ischemic/reperfused rat livers. Shock 1(1):3–9
2. Clemens MG, Bauer M, Pannen BH, et al (1997) Remodeling of hepatic microvascular responsiveness after ischemia/reperfusion. Shock 8(2):80–85
3. Clemens MG, Zhang JX (1999) Regulation of sinusoidal perfusion: in vivo methodology and control by endothelins. Semin Liver Dis 19(4):383–396
4. Zhang JX, Pegoli W Jr, Clemens MG (1994) Endothelin-1 induces direct constriction of hepatic sinusoids. Am J Physiol 266(4 Pt 1):G624–G632
5. Zhang JX, Bauer M, Clemens MG (1995) Vessel- and target cell-specific actions of endothelin-1 and endothelin-3 in rat liver. Am J Physiol 269(2 Pt 1):G269–G277
6. Bauer M, Bauer I, Sonin NV, et al (2000) Functional significance of endothelin B receptors in mediating sinusoidal and extrasinusoidal effects of endothelins in the intact rat liver. Hepatology 31(4):937–947
7. Rockey DC, Housset CN, Friedman SL (1993) Activation-dependent contractility of rat hepatic lipocytes in culture and in vivo. J Clin Invest 92(4):1795–1804
8. Pannen BH, Bauer M, Zhang JX, et al (1996) Endotoxin pretreatment enhances portal venous contractile response to endothelin-1. Am J Physiol 270(1 Pt 2):H7–H15
9. Pannen BH, Bauer M, Nolde-Schomburg GF, et al (1997) Regulation of hepatic blood flow during resuscitation from hemorrhagic shock: role of NO and endothelins. Am J Physiol 272(6 Pt 2):H2736–H2745
10. Sonin NV, Garcia-Pagan JC, Nakanishi K, et al (1999) Patterns of vasoregulatory gene expression in the liver response to ischemia/reperfusion and endotoxemia. Shock 11(3):175–179
11. Pannen BH, Bauer M, Zhang JX, et al (1996) A time-dependent balance between endothelins and nitric oxide regulating portal resistance after endotoxin. Am J Physiol 271(5 Pt 2):H1953–H1961
12. Bauer M, Paquette MC, Zhang JX, et al (1995) Chronic ethanol consumption increases hepatic sinusoidal contractile response to endothelin-1 in the rat. Hepatology 22(5): 1565–1576
13. Bauer I, Bauer M, Pannen BH, et al (1995) Chronic ethanol consumption exacerbates liver injury following hemorrhagic shock: role of sinusoidal perfusion failure. Shock 4(5):324–331
14. Yokoyama Y, Baveja R, Sonin N, et al (2000) Altered endothelin receptor subtype expression in hepatic injury after ischemia/reperfusion. Shock 13(1):72–78
15. Shah V, Haddad FG, Garcia-Cardena G, et al (1997) Liver sinusoidal endothelial cells are responsible for nitric oxide modulation of resistance in the hepatic sinusoids. J Clin Invest 100(11):2923–2930
16. Shah V, Toruner M, Haddad F, et al (1999) Impaired endothelial nitric oxide synthase activity associated with enhanced caveolin binding in experimental cirrhosis in the rat. Gastroenterology 117(5):1222–1228

Application of Superparamagnetic Iron Oxide for Hepatic Tumor Diagnosis

AKIHIRO TANIMOTO and SACHIO KURIBAYASHI

Summary. Superparamagnetic iron oxide (SPIO) particles as magnetic resonance (MR) contrast media are composed of iron oxide crystals coated with dextran or carboxydextran. These particles are sequestered by phagocytic Kupffer cells in normal reticuloendothelial system (RES), but are not retained in tumor tissue. Consequently, a significant differences in T2/T2* relaxation between normal RES tissue and tumors occurs, resulting in increased lesion conspicuity and detectability. The initial introduction of SPIO was expected to substantially increase the detectability of malignant hepatic tumors. It has been documented that SPIO-enhanced MR imaging is at least as accurate as CT during arterial portography in the detection of hepatic metastases, and is slightly better diagnostic performance than dynamic helical CT in the detection of hypervascular hepatocellular carcinomas. A combination of dynamic and static MR imaging technique using T1 and T2 imaging criteria appears to provide clinically more useful enhancement patterns. T2-weighted SPIO-enhanced MR imaging also provides useful clinical information by tumor enhancement of frequent benign tumors such as hemangiomas and RES-containing tumors compared to non-uptake of liver metastases. The possibility of one-step diagnosis is an attractive alternative to existing multi-step diagnoses in liver imaging, and is expected to be economically favorable.

Key words. Liver, Contrast media, Superparamagnetic iron oxide, Magnetic resonance imaging, Neoplasm

---

Department of Diagnostic Radiology, Keio University School of Medicine, 35 Shinanomachi, Shinjuku-ku, Tokyo 160-8582, Japan

Introduction

Because of the unique physiologic properties of the liver, improved depiction and characterization of focal hepatic lesions is possible with liver-specific magnetic resonance (MR) contrast agents. Superparamagnetic iron oxide (SPIO) particles are opsonized and sequestered by phagocytic Kupffer cells of normal reticuloendothelial system (RES). Phagocytosed SPIO particles in KCs produce strong T1, T2, and T2* relaxation effects of the liver parenchyma. Malignant tumors retain no KCs and show no signal change, resulting in increased tumor–liver contrast, which can be exploited to decrease threshold size for lesion detection.

Superparamagnetic iron oxide has been widely used for liver MR imaging for several years recently [1–3]. The appropriate use of tissue-specific agents would allow accurate detection and characterization of focal hepatic lesions. Ferumoxides and ferucarbotran are now commercially available as SPIO particles, and the latter is a bolus-injectable agent. In this chapter, current clinical evaluation of SPIO is reviewed with the literature and our experiences.

Superparamagnetic Iron Oxide Particles

A SPIO particle is a conglomerate of numerous iron oxide crystals coated with dextran or carboxydextran. The mean size of SPIO ranges between approximately 60 and 250 nm, which is certainly pertinent for the phagocytosis by Kupffer cells. Sequestered SPIO particles are metabolically biodegradable and bioavailable, being rapidly turned over into body iron stores and incorporated into erythrocyte hemoglobin. Two SPIO particle formulations are at present clinically available, ferumoxides and ferucarbotran (Table 1).

Superparamagnetism is a property intermediate to those of paramagnetic and ferromagnetic materials. Superparamagnetic materials comprise crystals of certain materials such as magnetite (Fe_3O_4) and maghemite (Fe_2O_3) large enough to form a solid phase microscopic volume or "domain" in which

TABLE 1. Features of commercially available SPIO particles

Commercial name	Feridex	Resovist
Generic name	Ferumoxides	Ferucarbotran
Particle size (nm)	100–250	57
r1 ($mM^{-1} s^{-1}$)	23	24
r2 ($mM^{-1} s^{-1}$)	100	168
Core material	$(Fe_2O_3)\ m \cdot (FeO)\ n$	$Fe_2O_3 + Fe_3O_4$
How to administer	Drip infusion	Bolus injection

atomic unpaired electron spins are aligned by positive exchange forces. When a domain is a volume of material (5–35 nm diameter for Fe_3O_4) possessing a uniform magnetization and a specific direction, superparamagnetism is displayed. Magnetic field gradients induced by superparamagnetic particles contribute to the dephasing of protons that move by diffusion in the vicinity of a particle, resulting in significant T2/T2* relaxation.

Clinical Applications

Despite the advent of newer treatment modalities, surgical resection is still considered the principal treatment for malignant focal hepatic lesions. In the preoperative evaluation of focal hepatic lesions, imaging methods that can reliably depict all malignant lesions are necessary, because imaging findings affect the choice of surgical intervention. Superparamagnetic iron oxide-enhanced MR imaging was more accurate than non-enhanced MR imaging and contrast-enhanced spiral computed tomography (CT) for the detection of focal hepatic lesions [2]. The combined analysis of non-enhanced and SPIO-enhanced images was more accurate in the characterization of focal hepatic lesions than was a review of SPIO-enhanced images alone [2].

Superparamagnetic iron oxide-enhanced MR imaging is particularly advantageous for detecting hepatic metastases, because the surrounding liver sustains normal phagocytic activity, and metastatic liver tumors have no Kupffer cells. Therefore, the diagnosis is simpler than hepatocellular carcinomas (HCC) derived from liver cirrhosis. It is superior to dual-phase CT and is equivalent to CT arterial portography (CTAP) [4,5] (Fig. 1). It has been documented that SPIO-enhanced MR imaging is more sensitive than dual-phase spiral CT, but is inferior to gadolinium-dynamic study for the depiction of hypervascular HCC [6–8]. However, SPIO offers additional information when imaging finding on dynamic MR imaging is questionable because of intrahepatic arterioportal shunt (AP shunt) and post-therapeutic liver damage.

Computed tomography arterial portography plus CT hepatic arteriography (CTHA) is a relatively invasive combination of modalities, but has been regarded as the most sensitive method for detecting focal hepatic lesions [9,10]. The combination of CTAP and CTHA is superior to CTAP alone for the detection of hypervascular HCC [10–11]. However, the use of CTAP plus CTHA is limited by pseudolesions due to intrahepatic AP shunt, and specificity was relatively low in the setting of chronic liver damage [11]. In our data (unpublished), the breath-hold SPIO-enhanced MR imaging protocol showed a diagnostic efficacy equivalent to that of the non-breath-hold MR imaging

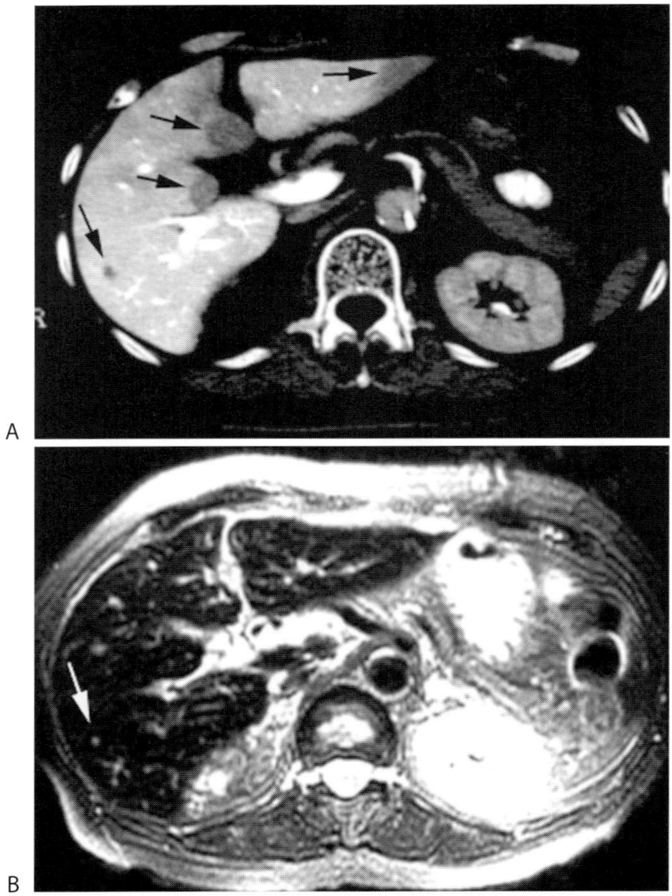

FIG. 1A,B. Seventy-nine-year-old female, liver metastasis and pseudolesions. **A** Computed tomography (CT) during arterial portography. Several perfusion defects are seen (*arrows*). **B** Superparamagnetic iron oxide (SPIO)-enhanced T2-weighted Fast Spin Echo (FSE: TR/TE = 2700/80 ms). Only one tiny lesion (diameter 3 mm) is noted (*arrow*). Other perfusion defects are proved to be pseudolesions

protocol and CTAP plus CTHA as a preoperative test for focal liver lesions (Table 2).

Some authors do not recommend CTAP plus CTHA for preoperative evaluation of HCC, because of the invasiveness, cost, and an unacceptably high false-positive rate without a substantial increase in sensitivity as compared with triple-phase helical CT [12]. Pseudolesions caused by AP shunt can be circumvented by the use of SPIO, since Kupffer cell function in liver parenchyma showing AP shunt is usually maintained [13]. It is well known that the inhomogeneity of SPIO uptake occurred because of reduc-

TABLE 2. Diagnostic efficacy of superparamagnetic iron oxide magnetic resonance imaging (SPIO-MRI)

	Protocol A	Protocol B	Protocol C
A. ROC analysis for 24 surgically proven metastases			
Az	0.96	0.96	0.95
Sensitivity (%)*	93.1	90.3	93.1
Specificity (%)**	99.4	99.4	95.8
Accuracy (%)**	98.3	97.8	95.3
B. ROC Analysis for 29 surgically proven HCC			
Az*	0.94	0.94	0.95
Sensitivity (%)*	88.5	89.7	90.8
Specificity (%)**	98.7	98.5	96.2
Accuracy (%)*	96.9	96.9	95.3

Protocol A, breath-hold SPIO-enhanced MR imaging; Protocol B, non-breath-hold SPIO-enhanced MR imaging plus Protocol A; Protocol C, computed tomography during arterial portography plus computed tomography during hepatic arteriography

ROC, receiver operating characteristic; Az, area under ROC curve; HCC, hepatocellular carcinomas

* Not significant; ** $P < 0.01$

tion in Kupffer cell density caused by inflammation, scarring, and regeneration in cirrhosis [14]. It should be noted that structural and functional inhomogeneity in cirrhosis could cause false positive lesions after SPIO administration.

The advantages of SPIO-enhanced MR imaging have been reported also in tissue characterization. Perfusion MR study using echo planar imaging allows a negative enhancement of hypervascular tumors [15], and one-stop shop diagnosis (both dynamic and RES-targeted MR imaging) for hypervascular HCC could be feasible in the future. Some types of hepatic lesions such as focal nodular hyperplasia (FNH), hepatocellular adenoma, adenomatous hyperplasia (AH), and well-differentiated HCC sustain phagocytic activity and may demonstrate iron oxide uptake [16,17]. The sustained phagocytic activity is a feature of FNH, which is helpful in differential diagnosis with SPIO-enhanced MR imaging [1,16] (Fig. 2). One criterion of a threshold signal loss of 10% on SPIO-enhanced MR images has been established to distinguish benign from malignant lesions (sensitivity 88%, specificity 89%) by receiver operating characteristic analysis [1].

Differentiation of HCC from AH is important for the early and precise detection of HCC in the cirrhotic liver. A study mentioned that there was no significant difference in the number of Kupffer cells between well-

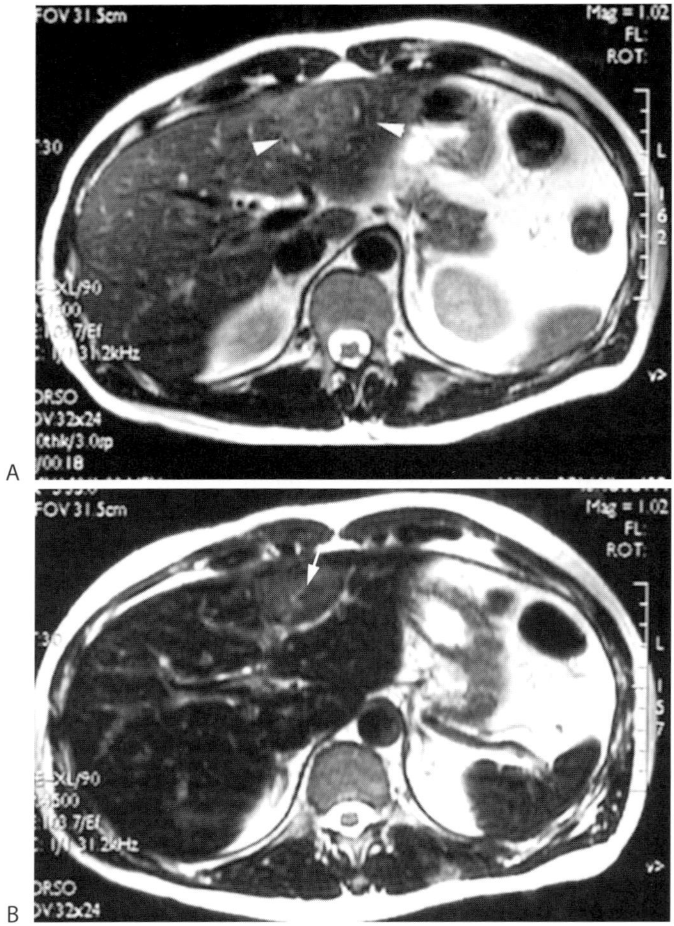

FIG. 2A,B. Forty-four-year-old male, focal nodular hyperplasia. **A** precontrast T2-weighted Fast SE (4500/103). A slightly hyperintense mass is noted in the lateral segment (*arrowheads*). **B** Post SPIO administration. The mass becomes slightly low intensity as compared with the precontrast image, suggesting sustained phagocytic activity. A central scar is also demonstrated more clearly (*arrow*)

differentiated HCC and surrounding liver tissue [17]. Therefore, it should be noted that phagocytic activity might overlap among borderline lesions.

The differentiation between hemangiomas and metastases is often difficult on the basis of their signal intensity on unenhanced, or gadolinium-enhanced MR imaging. It is known that SPIOs show a stronger T1 relaxation effect in the liver than do gadolinium-chelating agents. The ability to distinguish hemangiomas from metastases is based on their respective enhancement on

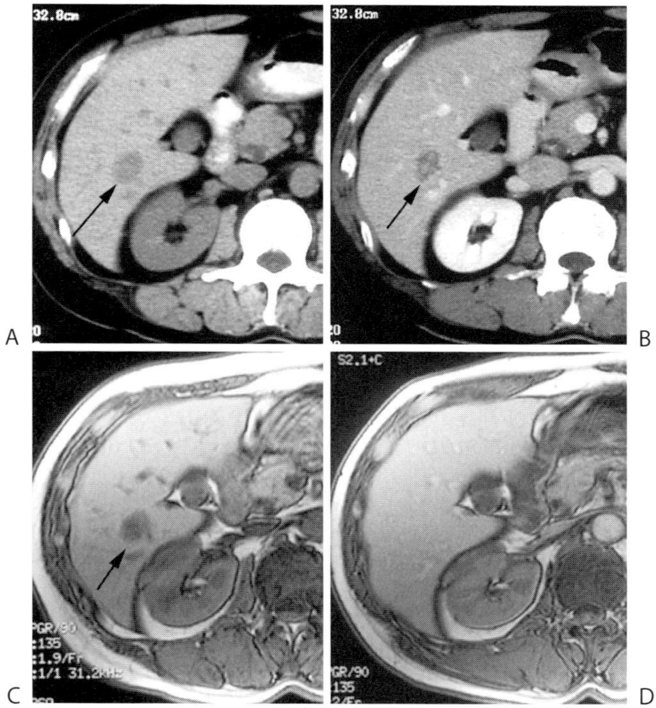

FIG. 3A–D. Fifty-two-year-old male, liver hemangioma. **A** Unenhanced CT. This patient who had rectal cancer underwent preoperative CT scan. A low-density mass is noted in S6 (*arrow*). **B** Contrast-enhanced CT. The mass still appears low density in delayed phase. Liver metastasis cannot be denied. **C** Unenhanced T1-weighted Fast spoiled GRASS (FSPGR 135/1.9/90°). A low-intensity mass is noted in S6. **D** Post SPIO administration. The liver signal intensity is slightly increased and the mass diminishes because of SPIO pooling in the vascular space of hemangioma

T1-weighted images [18] (Fig. 3). The feature for the differentiation between hemangiomas and metastases lies in the combination of the T1 blood pool effect, which positively enhances hemangiomas, and the T2 effect which negatively enhances the surrounding liver [19].

Conclusion

Superparamagnetic iron oxide-enhanced MR imaging is an effective imaging tool for the pretherapeutic evaluation and follow-up diagnostics of liver tumors with improved detection and differential information. Superparamagnetic iron oxide-enhanced MR imaging helps to improve the selection of patients who are candidates for curative liver surgery because invasive surgery will be avoidable if multiple lesions are present. With the exploitation of fast pulse sequences, the sensitivity of this technique will substantially increase. In view of comprehensive medical care, one-stop diagnosis with SPIO will be an attractive alternative to existing multi-modality based decision-making.

References

1. Vogl TJ, Hammerstingl R, Schwartz W, et al (1996) Superparamagnetic iron oxide-enhanced versus gadolinium-enhanced MR imaging for differential diagnosis of focal liver lesions. Radiology 198: 881–887
2. Reimer P, Jahnke N, Fiebich M, et al (2000) Hepatic lesion detection and characterization: value of non-enhanced MR imaging, superparamagnetic iron oxide-enhanced MR imaging, and spiral CT-ROC analysis. Radiology 217:152–158
3. Poeckler-Schoeniger C, Koepke J, Gueckel F, et al (1999) MRI with superparamagnetic iron oxide: efficacy in the detection and characterization of focal hepatic lesions. Magn Reson Imaging 17: 383–392
4. Haider MA, Amitai MM, Rappaport DC, et al (2002) Multi-detector row helical CT in preoperative assessment of small (< or = 1.5 cm) liver metastases: is thinner collimation better? Radiology 225:137–142
5. Senéterre E, Taourel P, Bouvier Y, et al (1996) Detection of hepatic metastases: ferumoxides-enhanced MR imaging versus unenhanced MR imaging and CT during arterial portography. Radiology 200:785–792
6. Lee JM, Kim IH, Kwak HS, et al (2003) Detection of small hypervascular hepatocellular carcinomas in cirrhotic patients: comparison of superparamagnetic iron oxide-enhanced MR imaging with dual-phase spiral CT. Korean J Radiol 4:1–8
7. Tang Y, Yamashita Y, Arakawa A, et al (1999) Detection of hepatocellular carcinoma arising in cirrhotic livers: comparison of gadolinium- and ferumoxides-enhanced MR imaging. AJR 172:1547–1554
8. Pauleit D, Textor J, Bachmann R, et al (2002) Hepatocellular carcinoma: detection with gadolinium- and ferumoxides-enhanced MR imaging of the liver. Radiology 222:73–80
9. Li L, Wu PH, Mo YX, et al (1999) CT arterial portography and CT hepatic arteriography in detection of micro liver cancer. World J Gastroenterol 5:225–227
10. Murakami T, Oi H, Hori M, et al (1997) Helical CT during arterial portography and hepatic arteriography for detecting hypervascular hepatocellular carcinoma. AJR 169:131–135
11. Makita O, Yamashita Y, Arakawa A, et al (2000) Diagnostic accuracy of helical CT arterial portography and CT hepatic arteriography for hypervascular hepatocellular carcinoma in chronic liver damage. An ROC analysis. Acta Radiol 41:464–469
12. Jang HJ, Lim JH, Lee SJ, et al (2000) Hepatocellular carcinoma: are combined CT during arterial portography and CT hepatic arteriography in addition to triple-phase helical CT all necessary for preoperative evaluation? Radiology 215:373–380
13. Oudkerk M, van den Heuvel AG, Wielopolski PA, et al (1997) Hepatic lesions: Detection with Ferumoxide-enhanced T1-weighted MR imaging. Radiology 203:449–456
14. Elizondo G, Weissleder R, Stark DD, et al (1990) Hepatic cirrhosis and hepatitis: MR imaging enhanced with superparamagnetic iron oxide. Radiology 174:797–801
15. Ichikawa T, Arbab AS, Araki T, et al (1999) Perfusion MR imaging with a superparamagnetic iron oxide using T2-weighted and susceptibility-sensitive echoplanar

sequences: evaluation of tumor vascularity in hepatocellular carcinoma. AJR 173: 207–213
16. Grandin C, Van Beers BE, Robert A, et al (1995) Benign hepatocellular tumors: MRI after superparamagnetic iron oxide administration. J Comput Assist Tomogr 19:412–418
17. Vogl TJ, Hammerstingl R, Schwartz W, et al (1996) Superparamagnetic iron oxide-enhanced versus gadolinium-enhanced MR imaging for differential diagnosis of focal liver lesions. Radiology 198:881–887
18. Tanaka M, Nakashima O, Wada Y, et al (1996) Pathomorphological study of Kupffer cells in hepatocellular carcinoma and hyperplastic lesions in the liver. Hepatology 24:807–812
19. Gansbeke VD, Metens TM, Matos C, et al (1997) Effects of AMI-25 on liver vessels and tumors on T1-weighted turbo-field-echo images: implications for tumor characterization. JMRI 7:482–489

Roles of Thromboxane in Lipopolysaccharide-Induced Hepatic Microcirculatory Dysfunction in Mice

Hiroyuki Katagiri[1], Yoshiya Ito[1], Ken-ichiro Ishii[1], Izumi Hayashi[2], Makoto Suematsu[3], Shuh Narumiya[4], Akira Kakita[1], and Masataka Majima[2]

Summary. Although thromboxanes (TXs) have been suggested to promote inflammation in the liver, little is known about the role of TXA_2 in leukocyte–endothelial interaction during endotoxemia. We observed using in vivo fluorescence microscopy that lipopolysaccharide (LPS) caused significant accumulation of leukocytes adhering to the hepatic microvessels and non-perfused sinusoids. Levels of serum alanine transaminase (ALT) and tumor necrosis factor α (TNFα) also increased. Lipopolysaccharide raised TXB_2 level in the perfusate from isolated perfused liver. A TXA_2 synthase inhibitor, OKY-046, and a TXA_2 receptor antagonist, S-1452, reduced LPS-induced hepatic microcirculatory dysfunction by inhibiting TNFα production. OKY-046 suppressed expression of intercellular adhesion molecule (ICAM)-1 in LPS-treated liver. In thromboxane prostanoid receptor-knockout mice, hepatic responses to LPS were minimized in comparison with those in their wild-type counterparts. These results suggest that TXA_2 is involved in LPS-induced hepatic microcirculatory dysfunction partly through the release of TNFα, and that endogenous TXA_2 could be responsible for the microcirculatory dysfunction during endotoxemia.

Key words. Thromboxane, Endothelial cell, Leukocyte, Tumor necrosis factor α, Adhesion molecule

---

[1] Department of Surgery, Kitasato University School of Medicine, 1-15-1 Kitasato, Sagamihara, Kanagawa 228-8555, Japan
[2] Department of Pharmacology, Kitasato University School of Medicine, 1-15-1 Kitasato, Sagamihara, Kanagawa 228-8555, Japan
[3] Department of Biochemistry and Integrative Medical Biology, Keio University, 35 Shinanomachi, Shinjuku-ku, Tokyo 160-8582, Japan
[4] Department of Pharmacology, Faculty of Medicine, Kyoto University, Yoshida, Sakyo-ku, Kyoto 606-8501, Japan

Introduction

The initial hepatic responses to lipopolysaccharide (LPS) include the activation of the nonparenchymal cells that constitute the hepatic microvascular system. The early events occurring in the hepatic microvasculature, including increases in leukocyte adhesion, reduction of sinusoidal perfusion, and activated Kupffer cells contribute to alterations in liver function caused by endotoxin [1]. However, the mechanisms by which LPS induces hepatic microcirculatory dysfunction are not fully understood.

Metabolites of arachidonic acid including prostaglandins (PGs) and thromboxanes (TXs) have been suggested to participate in liver injury during endotoxemia [2]. A significant increase in the plasma level of TXB_2 (a stable metabolite of TXA_2) is shown following LPS administration [3], and TXA_2 receptor antagonist exerts a protective effect on liver injury caused by endotoxin [4]. Furthermore, PGs and TXs modulate TNFα synthesis; PGE_2 suppresses TNFα production from Kupffer cells stimulated with endotoxin [5], while TXA_2 synthase inhibitor suppresses TNFα release from peritoneal macrophages [3]. These results suggest that TXA_2 could augment leukocyte-endothelial interaction during endotoxemia by affecting the production of TNFα.

The present study was thus conducted to examine the effects of the inhibition of TXA_2 synthase and of the blockade of TXA_2 receptor on the hepatic microvascular response to LPS in mice using in vivo microscopic methods. Some of the experiments were performed with thromboxane prostanoid (TP)-receptor knockout mice to elucidate the role of endogenously produced TXA_2 in this response.

Materials and Methods

Experimental Protocols for In Vivo Microscopic Study

Male C57BL/6 mice (6–8 weeks of age), weighing 20–25 g, were obtained from Shizuoka Laboratory Animal Center (Hamamatsu, Japan). TP receptor-knockout mice ($TP^{-/-}$, male, 8 weeks of age) were developed by us. All animals were provided food and water ad libitum. All procedures on animals were performed in accordance with the guideline for animal experiment of Kitasato University School of Medicine. Lipopolysaccharide was injected intravenously (0.3 mg/kg in 0.1 ml of physiological saline) into mice through the tail vein. OKY-046, a thromboxane synthase inhibitor (50 mg/kg in 0.1 ml saline, i.v.), S-1452, a TP receptor antagonist (10 mg/kg, p.o.), and vehicle (5% gum arabic, 0.1 ml/mouse, p.o.) were administered 30 min before LPS injection. Anti-LFA-1 monoclonal antibody was administered (2 mg/kg, i.v.) simultaneously with LPS injection.

Preparation for In Vivo Microscopy

Four hours after LPS injection, animals anesthetized with pentobarbital sodium (50 mg/kg, i.p.) were prepared for in vivo fluorescence microscopy according to the methods previously described [6]. The number of leukocytes adhering was determined off-line during video playback analysis. A leukocyte was defined as adhering to the venular and sinusoidal walls if it remained stationary for more than 20 s. With respect to the leukocytes adhering to the venules, 5–8 portal or central venules per animal were observed and assessed. The endothelial surface area of each venule was measured from video recordings using an adjustable electric microscaler (Argus-10; Hamamatsu Photonics; Hamamatsu, Japan). We estimated the adhesion of leukocytes in terms of (1) number of adhered leukocytes per observation microscopic field (×200) in sinusoids, and (2) number of adhered leukocytes per 1000 μm^2 of endothelial surface in portal venules and central venules.

The sinusoidal perfusion deficits were evaluated by counting the number of nonperfused sinusoids in the same microscopic field as that in which the number of adhering leukocytes was determined. The percentage of nonperfused sinusoids was calculated as the ratio of the number of nonperfused sinusoids to the total number of all visible sinusoids. The results were expressed as the percentage of nonperfused sinusoids.

Four hours after LPS injection, approximately 100 mg of the liver tissue was excised. Reverse transcription–polymerase chain reaction (RT-PCR) and immunohistochemistry were performed according to our previous report [7].

Experimental Procedure for Perfusion of the Liver

The isolated nonrecirculating perfused liver system was prepared according to the method of Suematsu et al. [8]. Four experimental groups of animal livers were set up to investigate whether the liver is the site of TXA_2 production. In the first group, the livers of wild-type mice perfused with buffer solution throughout the experiment served as controls. In the second and third groups, 30 min after the start of perfusion, the administration of LPS (1.25 μg/min for 20 min) was initiated to TP-receptor knockout mice and to their wild-type counterparts and was continued throughout the experimental period. In the fourth group, 30 min before the start of preparation for the perfusion experiment, wild-type mice were treated with OKY-046 (50 mg/kg, i.v.), and OKY-046 (0.05 μg/min for 50 min) was administered simultaneously with the start of the perfusion with buffer. At 30 min after the start of perfusion, LPS (1.25 μg/min for 20 min) was continuously infused until the end of the experimental period.

Results and Discussion

The administration of LPS caused significant increases in the numbers of leukocytes adhering to the portal venules (8.5-fold) (Fig. 1A), sinusoids (50.2-fold) (Fig. 1B), and central venules (51.0-fold) (Fig. 1C) in comparison with those in saline-treated mice. Pretreatment with OKY-046 lowered those by 61%, 46%, and 45%, respectively. Pretreatment with S-1452 also suppressed those by 69%, 48%, and 39%, respectively. Concomitantly, the percentage of nonperfused sinusoids after LPS injection was increased (7.1-fold) (Fig. 1D). The percentage of nonperfused sinusoids was significantly lowered by OKY-046 (by 61%) and S-1452 (by 47%), respectively.

OKY-046 decreased the levels of plasma alanine transaminase (ALT) activity at 4h after, and plasma TNFα at 1h after, LPS injection by 22% and 31%, respectively. S-1452 also reduced them by 17% and 40%, respectively (Fig. 2).

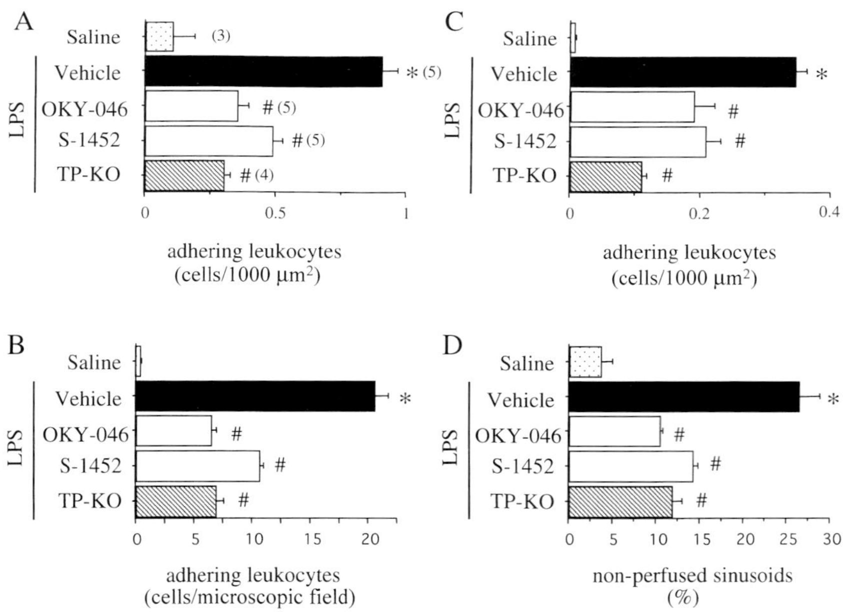

FIG. 1. Effects of lipopolysaccharide (*LPS*) on (**A**) the numbers of leukocytes adhering to the portal venules, (**B**) the numbers of leukocytes adhering to the sinusoids, (**C**) the numbers of leukocytes adhering to the central venules, and (**D**) the percentage of non-perfused sinusoids, all in thromboxane prostanoid receptor-knockout (*TP-KO*) mice and in their wild-type counterparts treated with OKY-046 (50 mg/kg, i.v.) and S-1452 (10 mg/kg, p.o.). Numbers in parentheses indicate number of animals. Data are shown as mean ± SEM. $^{*}P < 0.05$ vs saline-treated mice; $^{\#}P < 0.05$ vs LPS-treated mice

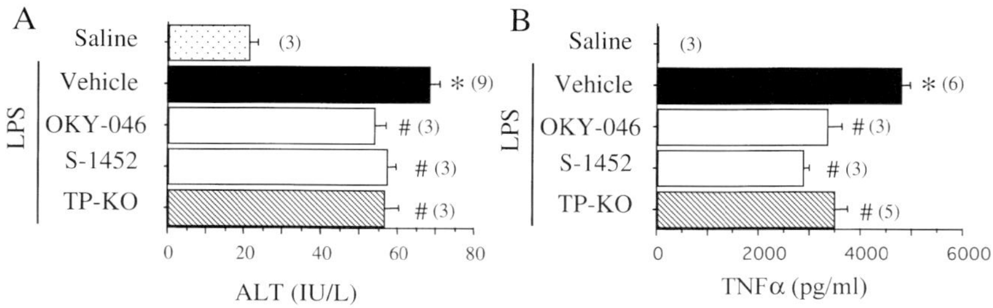

FIG. 2. Effects of LPS on (**A**) serum alanine aminotransferase (*ALT*) activity and (**B**) the serum concentrations of tumor necrosis factor α (*TNFα*) in TP receptor-knockout (*TP-KO*) mice and in their wild-type counterparts treated with OKY-046 (50 mg/kg, i.v.) and S-1452 (10 mg/kg, p.o.). Numbers in parentheses indicate number of animals. Data are shown as mean ± SEM. $^{*}P < 0.05$ vs saline-treated mice; $^{\#}P < 0.05$ vs LPS-treated mice

To further investigate whether LPS-induced hepatic microcirculatory dysfunction is mediated by endogenously produced TXs, we used TP receptor-knockout mice. In wild-type counterparts, LPS caused significant hepatic microcirculatory dysfunction as described above (Fig. 1A–D). In TP receptor-knockout mice, the numbers of leukocytes adhering to the portal venules, sinusoids, and central venules were significantly lower than in wild-type counterparts. Also the percentage of nonperfused sinusoids was lower than wild-type mice. The levels of ALT and TNFα following LPS administration in TP receptor-knockout mice were decreased by 18% and 28%, respectively (Fig. 2).

Figure 3 illustrates changes in the levels of TXB_2 in the effluent perfusate from isolated perfused liver. The perfusion experiments were performed to determine the site of generation of TXA_2. During the isolation procedure of the liver, we used heparin. After set up of the isolated liver, we perfused LPS to see the release of TXA_2. We used KH buffer as vehicle solution of LPS perfusion. In both cases, we used heparin during isolation procedure. In controls, no significant change in TXB_2 levels appeared (Fig. 3A). Within 15 min of the start of LPS administration, TXB_2 levels were rapidly increased in comparison with the baseline, and then continued to increase (Fig. 3B). Treatment of wild-type mice with OKY-046 completely abolished the increment of TXB_2 in response to LPS (Fig. 3C). In TP receptor-knockout mice, changes in TXB_2 levels after LPS were similar to those in the wild-type counterparts (Fig. 3D). During the perfusion experiment, the perfusion pressure was stable (2–4 cmH_2O) at all.

To investigate whether attenuation of hepatic microcirculatory dysfunction by TXA_2 inhibition affected the expression of adhesion molecules, the expression of ICAM-1 and platelet endothelial cell adhesion molecule (PECAM)-1

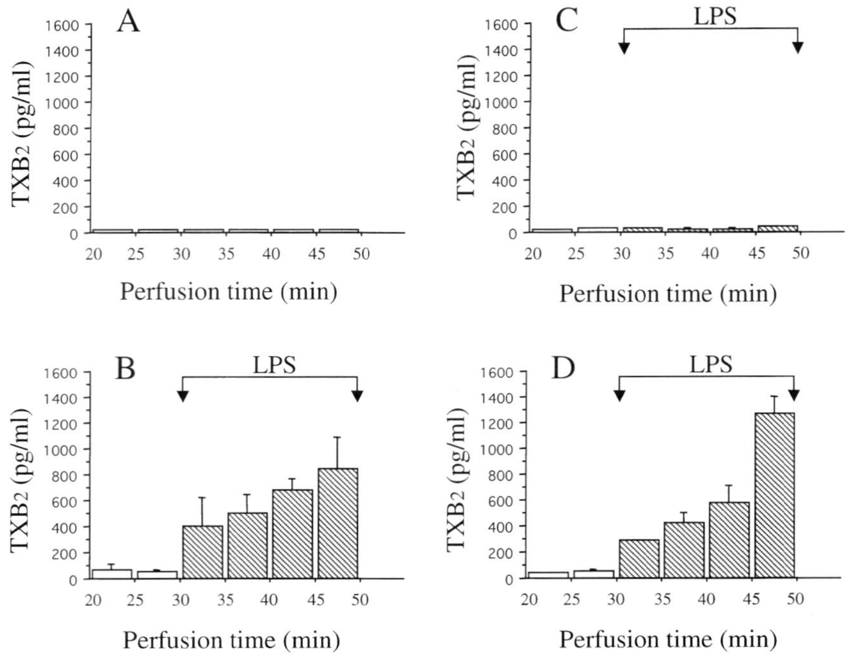

FIG. 3. Changes in the levels of thromboxane B_2 (*TXB_2*) in the effluent perfusate from the isolated perfused livers of wild-type mice (**A–C**) and of TP receptor-knockout mice (**D**). Livers from LPS-treated wild-type mice (**B**) and of TP receptor-knockout mice (**D**). Livers perfused with KH buffer solution alone throughout the experiment served as controls (**A**). Livers from wild-type mice treated with a combination of OKY-046 (0.05 μg/min for 50 min) and LPS (1.25 μg/min for 20 min) (**C**). Data are shown as mean ± SEM from three animals. *$P < 0.05$

in the liver was assessed by RT-PCR and by immunohistochemistry. Lipopolysaccharide resulted in enhanced hepatic expression of mRNA of ICAM-1 when compared with that in saline-treated mice. OKY-046 reduced the expression of ICAM-1 mRNA. These reductions were seen with TP-KO mouse liver as well. The immunoreactivity with ICAM-1 was demonstrated in the sinusoids of saline-treated mice. Lipopolysaccharide upregulated ICAM-1 expression in the sinusoids as well as the hepatic venules, and OKY-046 reduced the ICAM-1 immunoreactivity.

To examine the significant contribution of ICAM-1 to leukocyte adhesion to the hepatic microvessels, we observed the effect of anti LFA-1 (lymphocyte function associated antigen 1, CD11a), the ligands of ICAM-1 on leukocyte, on hepatic microcirculation in vivo microscopic study. Anti LFA-1 antibody significantly lowered the numbers of leukocytes adhering to the portal venules, sinusoids, and central venules, respectively. The percentages of nonperfused sinusoids also were lowered by anti LFA-1.

Lipopolysaccharide resulted in slightly enhanced hepatic expression of mRNA of PECAM-1 when compared with that in saline-treated mice; however, OKY-046 did not affect the expression of PECAM-1 in response to LPS. In the immunohistochemical study, PECAM-1 was demonstrated in saline-treated mouse liver. Lipopolysaccharide enhanced expression of PECAM-1 weakly in comparison with that of ICAM-1. OKY-046 did not affect PECAM-1 expression induced by LPS.

The results of the present study showed that OKY-046, a TXA_2 synthase inhibitor and S-1452, a TXA_2 receptor antagonist, attenuated LPS-induced hepatic microcirculatory dysfunction including an increase of leukocytes adhering to the hepatic microvessels, as well as impaired sinusoidal perfusion. These reductions were accompanied by decreases in the serum levels of ALT and TNFα. To rule out the possibility that endotoxin may changes systemic hemodynamics, we measured arterial blood pressure. A dose of LPS (0.3 mg/kg, i.v.) used in the present study did not reduce arterial blood pressure. These results suggest that TXA_2 enhances hepatic microcirculatory dysfunction during endotoxemia. This possibility was supported by our finding that TP receptor-knockout mice minimized liver injury and hepatic microcirculatory dysfunction in response to LPS by inhibiting TNFα production. Our finding that TXA_2 appears to modulate TNFα production following LPS administration (Fig. 2A) is consistent with the findings of others [3,9–11] that TXA_2 synthase inhibitor and TXA_2 receptor antagonist decrease TNFα levels. The inhibitory effects of OKY-046 and S-1452 on TNFα production in response to LPS were partial (30%–40% reduction), while OKY-046 completely inhibited TXB_2 release in perfusate from LPS-treated liver. These results suggested that other factors may be involved in the regulation of TNFα generation.

Leukocyte adhesion to the hepatic microvessels seems to be critical in microcirculatory dysfunction. It was reported that TXA_2 receptor antagonist completely suppressed the expression of ICAM-1 on human umbilical vein endothelial cells [12]. We showed that OKY-046 suppressed the expression of ICAM-1. Thus attenuation of hepatic leukocyte adhesion is attributed to inhibition of ICAM-1 expression by TXA_2 inhibition. Furthermore, the current study showed that PECAM-1 seems to be slightly increased in LPS-treated livers and OKY-046 did not inhibit its expression. It suggested that PECAM-1 did not exhibit a major role in LPS induced hepatic microcirculatory dysfunction. However, ICAM-1 was intensely increased by LPS and was suppressed by TXA_2 blockade. We also examined effect of anti LFA-1, ligand of ICAM-1, in hepatic microcirculatory dysfunction. In result, anti LFA-1 improved the hepatic microcirculatory dysfunction, suggested that ICAM-1 induced leukocyte adhesion and sinusoidal perfusion deficits functionally.

Lipopolysaccharide administration to the isolated perfused liver resulted in a rapid and significant release of TXB_2 into the perfusate, suggesting that the liver is an important source of TXA_2, and that TXA_2 seems to be an inflammatory mediator in the early phase of endotoxemia. We measured the levels of TXB_2 in the effluent perfusate, because the levels of TXs and PGs were artificially elevated with ease by mechanical stimulation. In the liver, TXA_2 in response to LPS is released from nonparenchymal cells, i.e., Kupffer cells [2] and sinusoidal endothelial cells [13]. Of these, Kupffer cells are a major source of TXA_2. However, the possibility that activated platelets are a productive source of TXA_2 in vivo cannot be excluded.

In conclusion, our present study clarifies that TXA_2 plays an important role in hepatic microcirculatory dysfunction elicited by LPS administration. Lipopolysaccharide-induced hepatic microcirculatory dysfunction was associated with TXA_2 generation in the liver, and TP receptor signaling is related to the upregulation of expression of an adhesion molecule, ICAM-1.

References

1. McCuskey RS (1993) Hepatic microvascular response to endotoxemia and sepsis. Prog Appl Microcirc 19:76–84
2. Decker K (1990) Biologically active products of stimulated liver macrophages. Eur J Biochem 192:245–361
3. Altavilla D Squadrito F, Canale P, et al (1995) G619, a dual thromboxane synthase inhibitor and thromboxane A_2 receptor antagonist, inhibits tumor necrosis factor-α biosynthesis. Eur J Pharmacol 286:31–39
4. Ishiguro S, Arii S, Monden K, et al (1994) Identification of the thromboxane A_2 receptor in hepatic sinusoidal endothelial cells and its role in endotoxin-induced liver injury in rats. Hepatology 20:1281–1286
5. Karck U, Peters T, Decker K (1988) The release of tumor necrosis factor from endotoxin-stimulated rat Kupffer cells is regulated by prostaglandin E_2 and dexamethasone. J Hepatol 7:352–361
6. Matsumoto Y, Ito Y, Hayashi I, et al (2002) Effect of FR167653, a novel inhibitor of tumor necrosis factor alpha and interleukin-1 beta synthesis on lipopolysaccharide-induced hepatic microvascular dysfunction in mice. Shock 17:411–415
7. Amano H, Hayashi I, Endo H, et al (2003) Host prostaglandin E(2)-EP3 signaling regulates tumor-associated angiogenesis and tumor growth. J Exp Med 197:221–232
8. Suematsu M, Goda N, Sano T, et al (1995) Carbon monoxide: an endogenous modulator of sinusoidal tone in the perfused rat liver. J Clin Invest 96:2431–2437
9. Kuhn DC, Stauffer JL, Gaydos LJ, et al (1993) An inhibitor of thromboxane production attenuates tumor necrosis factor release by activated human alveolar macrophages. Prostaglandins 46:195–205
10. Grandel U, Fink L, Blum A, et al (2000) Endotoxin-induced myocardial tumor necrosis factor-α synthesis depresses contractility of isolated rat hearts. Evidence for a role

of sphingosine and cyclooxygenase-2-derived thromboxane production. Circulation 102:2758–2764
11. Caughey GE, Pouliot M, Cleland L, et al Regulation of tumor necrosis factor-α and IL-1 synthesis by thromboxane A_2 in nonadherent human monocytes. J Immunol 158:351–358
12. Ishizuka T, Kawakami M, Hidaka T, et al (1998) Stimulation with thromboxane A_2 (TXA_2) receptor agonist enhances ICAM-1, VCAM-1 or ELAM-1 expression by human vascular endothelial cells. Clin Exp Immunol 112:464–470
13. Rieder H, Ramadori G, Allmann K-H, et al (1990) Prostanoid release of cultured liver sinusoidal endothelial cells in response to endotoxin and tumor necrosis factor. J Hepatol 11:359–366

Changes of Hepatic Microcirculation in Acute Cholestasis

Takashi Osada[1], Tetsuo Morishita[1], Eiichi Sekizuka[2], Kouji Miyazaki[2], Eisuke Iwasaki[1], Hiroshi Kishikawa[1], Masaru Nakano[1], Tadashi Ohara[1,3], Jiro Nishida[1], Hiroshi Nagata[4], and Hiromasa Ishii[4]

Key words. Microcirculation, Liver, Cholestasis, Bile duct pressure, Portal vein

Introduction

Hepatocellular dysfunction is often seen in cholestasis and biliary obstruction such as cholelithiasis and choledocholithiasis, and after endoscopic retrograde cholangiopancreatography (ERCP). The purpose of this study was to investigate hepatic microcirculatory changes in the increase of bile duct pressure experimentally.

Materials and Methods

Male Wistar rats (200–300 g) fasted overnight were anesthetized with 50 mg/kg sodium pentobarbital i.p. The common bile duct was cannulated with a Clear Vinyl Tube (o.d. 0.8 mm) and pressure was continuously measured with a Pressure Monitor, PB-1 (World Instruments).

Study 1

The pressure of the bile duct was increased from basal level (+0 cmH_2O) (13.7 ± 2.2 cmH_2O, mean ± SE, $n = 7$) to +10 cmH_2O, +20 cmH_2O, and +30 cmH_2O, and kept for 6 min at each step. Systemic blood pressure was monitored

---

[1] Department of Internal Medicine and Gastroenterology, Tokyo Dental College, Ichikawa General Hospital, 5-1-13 Sugano, Ichikawa, Chiba 272-8513, Japan
[2] Saitama National Hospital, 2 Suwa, Wako, Saitama 351-0102, Japan
[3] Tokyo Dental Collage, Chiba Hospital, 1-1-2 Masago, Mihama-ku, Chiba 261-8502, Japan
[4] Department of Internal Medicine, Keio University School of Medicine, 35 Shinanomachi, Shinjuku-ku, Tokyo 160-8582 Japan

throughout the experiments. Blood flow at the surface of the liver was measured with a laser Doppler flowmeter, ALF2000 (Advance, Tokyo, Japan). Blood flow in the portal trunk was measured with an Electromagnetic Blood flowmeter, MVF-3200 (Nippon Kohden, Tokyo, Japan). Microcirculation of the liver was observed using in vivo microscopy [1]. The velocity of red blood cells labeled with fluorescein isothiocyanate (FITC) in the hepatic sinusoids was measured using a high-speed video system [2] with Ektapro HS-4500 (Redlake MASD, San Diego, CA, USA).

Study 2

Rats were divided into two groups with and without increase of the bile duct pressure in the same way as study 1. Serum levels of TB, AST, ALT, LDH, ALP, GGT, and LAP were measured, and the liver was examined microscopically in both groups.

The Friedman test, Bonferroni test and Mann–Whitney *U*-test were used for statistical analysis. All values are expressed as mean ± SE, with $P < 0.05$ accepted as significant.

Results

Study 1

The mean systemic pressure (132 ± 4 mmHg at the basal level, $n = 10$) did not change throughout the experiment. The hepatic blood flow decreased significantly ($P < 0.05$) from 100% ± 0% (11.8 ± 2.9 ml/min per 100 g, $n = 4$) at +0 cmH_2O, 91% ± 2% at +10 cmH_2O, 75% ± 9% at +20 cmH_2O, to 54% ± 7% at +30 cmH_2O (Fig. 1) during the stepwise increase of the bile duct pressure. The blood flows in the portal trunk and blood cell velocity in the hepatic sinusoids also decreased significantly.

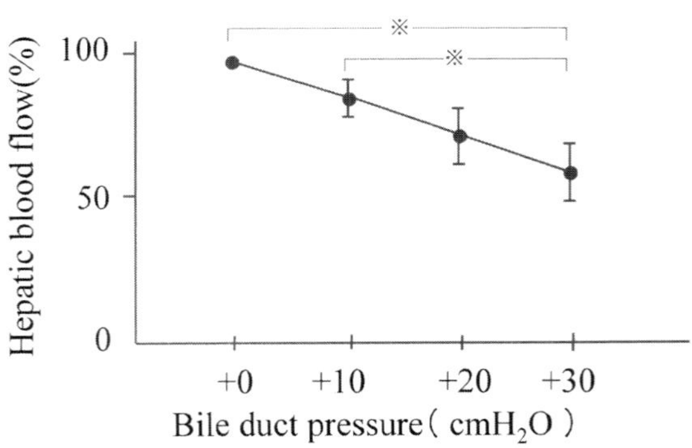

FIG. 1. Hepatic blood flow in the increase of bile duct pressure. The hepatic blood flow decreased significantly (*$P < 0.05$) from 100% ± 0% (mean ± SE 11.8 ± 2.9 ml/min per 100 g, $n = 4$) at +0 cmH_2O during the stepwise increase of bile duct pressure

Study 2

There was no significant difference in the serum levels of TB, AST, ALT, LDH, ALP, GGT, and LAP between both groups. Microscopic examination did not show abnormal histological findings of the liver such as bile thrombus formation in the bile canaliculi, and degeneration of hepatocytes in either of the groups.

Discussion

Our study showed that increase of the bile duct pressure decreased blood flow on the liver surface and in the portal trunk, and blood cell velocity in the hepatic sinusoids without significant change of the blood chemistry and abnormal histological findings of the liver. This suggests that hepatic microcirculatory changes are the previous step of hepatocellular dysfunction in biliary obstruction. Doppman et al. [3] reported that the liver with biliary obstruction is easily associated with infarction or abscess formation by hepatic artery ligation or embolization. The change of the hepatic blood flow might affect the prognosis of obstructive jaundice [4]. Further study on hepatic arterioles is needed.

Conclusion

The hepatic blood flow decreases with the decrease of blood cell velocity in the sinusoids and portal blood flow in acute cholestasis. This microcirculatory change precedes the biochemical and histological changes in the liver.

References

1. Morishita T (1997) Microcirculation of the gastrointestinal tract in acute portal hypertension. Microcirc Annu 13:39–42
2. Ueki I, Sekizuka E, Okada E, et al (1993) Measurement of red blood cell velocity vector with a combination of a real-time and simultaneous multiwindows cross-correlation and high-speed video system. Bioimages 1:149–157
3. Kodama O, Fujii Y, Tanaka T, et al (1990) Experimental and clinical studies on liver tissue blood flow in obstructive jaundice. Nippon Shokakibyo Gakkai Zasshi 87:49–56
4. Doppman JL, Girton M, Vermess M (1982) The risk of hepatic artery embolization in the presence of obstructive jaundice. Radiology 143:37–43

Part 4
Cell Adhesion and Traffic in Microcirculation

Microvascular Aspects of Ischemia–Reperfusion Injury

Thiruma V. Arumugam and D. Neil Granger

Summary. Ischemia–reperfusion (I/R) has been implicated in the pathogenesis of a number of diseases that affect a variety of organ systems. Research on this problem has led to the recognition that the microcirculation is particularly vulnerable to the deleterious effects of I/R. The microvascular dysfunction caused by I/R is manifested as impaired endothelium-dependent dilation in arterioles, enhanced fluid filtration and leukocyte plugging in capillaries, and the trafficking of leukocytes and platelets, as well as extravasation of plasma proteins in postcapillary venules. Activated endothelial cells in all segments of the microcirculation produce more oxygen radicals, and the bioavailability of nitric oxide diminishes after reperfusion. The resulting imbalance between superoxide and nitric oxide in endothelial cells leads to the production and release of inflammatory mediators and an enhanced biosynthesis of and increased cell surface expression of adhesion molecules that mediate the recruitment of both leukocytes and platelets. More recent evidence indicates that I/R reveals a link between inflammation and hemostasis, such that the accumulation of leukocytes coincides with the recruitment of platelets. Our work on intestinal venules indicates that approximately 40% of the adherent leukocytes are platelet bearing, and that platelets utilize P-selection to bind to the P-selectin glycoprotein ligand-1 that is constitutively expressed on the surface of leukocytes. A fraction of the platelets that accumulate in postischemic venules bind directly to activated endothelial cells. Some of the known risk factors for cardiovascular disease, including hypercholesterolemia, hypertension, and diabetes appear to exaggerate many of the microvascular alterations elicited by I/R. Finally, the inflammatory mediators released as a consequence of reperfusion also appear to activate endothelial

---

Department of Molecular and Cellular Physiology, Louisiana State University Health Sciences Center, Shreveport, LA 71130-3932, USA cells in remote organs that are not directly exposed to the initial ischemic insult. This distant response to I/R can result in leukocyte-dependent microvascular injury that is characteristic of the multiple organ dysfunction syndrome.

Key words. Ischemia–reperfusion, Oxidative stress, Leukocyte–endothelial cell adhesion, Platelet adhesion

Introduction

Ischemia is a marked reduction in the blood flow to an organ or vascular bed. Ischemia can lead to significant tissue injury and cell death if it is prolonged. Reperfusion injury can be defined as the damage that occurs to an organ during the resumption of blood flow following an episode of ischemia. This can be distinguished from the injury caused by ischemia, although the conditions needed to cause a reperfusion injury are generated by the ischemic episode. Reperfusion injury has a complex pathophysiology, and involves an orchestrated sequence of cellular and molecular events. Prolonged ischemia can result in endothelial dysfunction and these responses to ischemia are exacerbated during reperfusion [1–4].

The microvascular response to I/R injury induces activation of xanthine oxidase [5], enhanced generation of superoxide and hydrogen peroxide [5,6], activation of nuclear transcription factors NF-κB and AP-1 [1,2,7], increased expression of endothelial cell adhesion molecules [8], increased adhesivity to leukocytes and platelets, and a reduction in endothelial barrier function [8–10]. Adhesion molecules on endothelial cells and leukocytes ensure an orderly sequence of leukocyte–endothelial cell interactions that promote leukocyte adherence to the endothelium and subsequent transendothelial migration into inflamed tissue [1]. There is growing evidence that also supports a role for platelets during the pathogenesis of reperfusion injury. Activated platelets are a source of various inflammatory mediators including arachidonic acid products, neutrophil activating peptide, and oxygen radicals [10–12]. Platelets express several adhesion molecules that mediate platelet adhesion to subendothelial matrix proteins [10,13]. Platelets have been shown to accumulate in the postischemic microvasculature early after reperfusion [14,15]. The objectives of this chapter are to summarize recent advances made in microvascular responses for I/R injury and outline mechanisms that are involved in the recruitment of leukocytes and platelets during I/R. This chapter also analyzes other factors that influence the I/R-induced microvascular dysfunction.

Microvascular Endothelial Responses for I/R Injury

The endothelial cells that line the inner surface of blood vessels are a vital and dynamic structure that is essential for vascular homeostasis. Prolonged ischemia alters membrane potential, disturbs the distribution of ions, increases intracellular volume, decreases membrane fluidity and impairs the cytoskeletal organization of endothelial cells [1,2,8,16]. These changes progress due to depletion of intracellular energy stores, a diminished production of certain bioactive agents such as prostacyclin, nitric oxide and an accelerated production of proinflammatory agents such as oxygen radicals, cytokines, complement products, arachidonic acid metabolites, and endothelin (Fig. 1) [1,16,17]. Endothelial cells become rounded, capillary lumina are narrowed and the vessels become leaky. Endothelial cells in all segments of the microcirculation are similarly exposed to the detrimental effects of I/R; however, the resultant endothelial cell dysfunction appears to be manifested in a site-specific manner.

Endothelial responses to I/R injury can be divided into arteriolar, capillary and venular responses. Arteriolar endothelial cells are necessary for agents like acetylcholine and bradykinin to cause vasodilation [18–20]. Under normal conditions, the vasoconstricting and vasodilating agents produced by endothelial cells are in dynamic balance and influence each other through multiple mechanisms [1,19,20]. Endothelial cells are in close proximity to underlying arteriolar smooth muscle and vasoactive agents produced by endothelium can cause profound changes in vascular tone [1–3,19,20]. Under normal conditions, both NO and superoxide anion are produced by endothelial cells. It is generally considered that the endothelium, particularly in the microvasculature, is the primary cellular source of free radicals in I/R injury. During the ischemic period, the enzyme xanthine dehydrogenase is converted to the oxidase isoform. At the same time, there is progressive breakdown of high-energy phosphates such as adenosine triphosphate. This leads to the accumulation of purine metabolites, xanthine, and hypoxanthine, which are substrates for xanthine oxidase [21–23]. Although the xanthine oxidase pathway is widely regarded as the dominant source of endothelial free radicals during reperfusion, it certainly is not the only one. Nicotinamide adenine dinucleotide phosphate (NADPH)-linked oxidase may be another major source of hydrogen peroxide in cells subjected to reoxygenation [24]. Normally, the production of NO by endothelial cells is 2–3 magnitudes higher than superoxide production and this allows NO to effectively scavenge intracellular superoxide anion, prevent platelet aggregation, and minimize adhesivity between endothelial cells and leukocytes [25]. Under I/R conditions the overproduction of superoxide anions by endothelial cells dramatically reduce NO bioavailability and NO synthase activity may be inhibited [1–3].

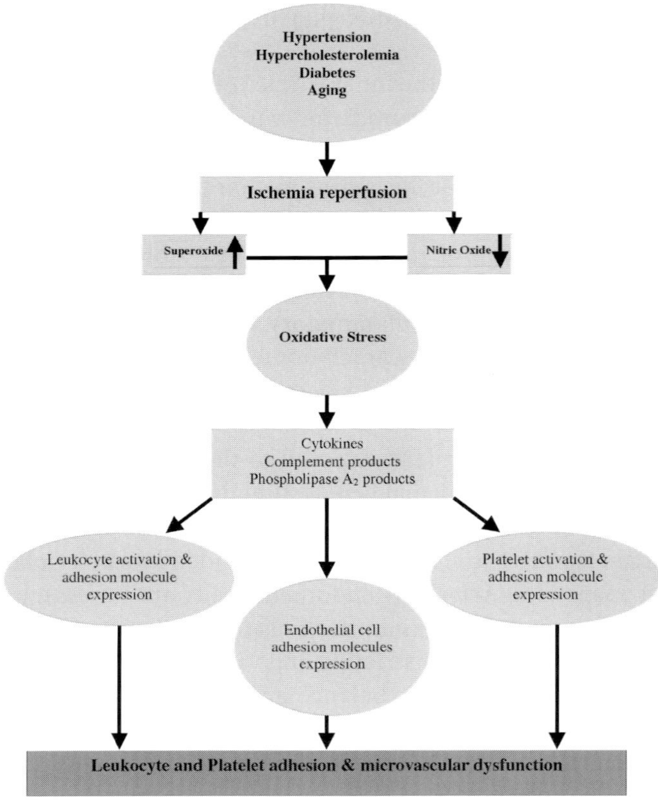

FIG. 1. Mechanisms underlying the microvascular dysfunction elicited by ischemia–reperfusion. Ischemia–reperfusion leads to an imbalance between the production of superoxide and nitric oxide. The resulting oxidative stress leads to the activation of proinflammatory agents such as cytokines, complement products, and phospholipase A2 products. As a consequence of these events, leukocytes and platelets adhere to microvascular endothelial cells, which leads to microvascular dysfunction

Overproduced superoxide anions and proinflammatory cytokines such as tumor necrosis factor-α, interleukin (IL)-1a, and IL-8 may account for the inability of arterioles to exhibit endothelium-dependent and NO-mediated vasodilation [26,27]. Although endothelial cells can generate these superoxide anions and proinflammatory cytokines during I/R period in the absence of leukocytes [29], leukocytes, particularly activated neutrophils, represent another potential source of these products. Mice deficient in leukocyte or endothelial cell adhesion molecules do not exhibit the impaired endothelium-dependent vasodilation normally seen after I/R [18].

Measurements of PO_2 in arterioles and capillaries during the reperfusion period have revealed that capillaries exhibit a lower and more variable PO_2

[28]. However, PO_2 in arterioles and tissue return to baseline values after 30 min of reperfusion [28]. This repeat along with other studies [29,30] indicates that anoxic zones always start at the capillaries during I/R and endothelial cells of the capillaries are extremely vulnerable to I/R. Capillary endothelial dysfunction results in increased filtration of fluid into the interstitium and a reduction in the number of perfused capillaries [1,31]. Capillary malperfusion has been described in many organ-specific I/R injuries [31–33]. Malperfusion of capillaries appears to result from the plugging of capillaries by stiffer, activated leukocytes with or without platelets [1,2,10]. Support for leukocyte-dependent mechanisms is provided by improved capillary perfusion in mice that are genetically deficient in leukocyte or endothelial adhesion molecules [34]. However, mechanisms independent of leukocyte adhesion and platelet aggregation have also been indicated in the capillary malperfusion in I/R [35]. Recent studies showed that endothelial cell and/or tissue swelling in combination with luminal obstruction and leukocyte plugging may be responsible for the early capillary malperfusion [35].

Activation of endothelial cells in postcapillary venules account for most of the inflammatory responses that are elicited by I/R. While endothelial cells in arterioles, capillaries, and venules all experience oxidant stress; those cells lining venules appear to bear the brunt of this response. This is due to a number of factors, such as oxygen radical production from both activated polymorphonuclear leukocytes and endothelial cells [1–3,7–9], release of proinflammatory substances from mast cells, macrophages situated in proximity to postcapillary venules [1,36], and reduced bioavailability of NO. Characteristic features of the venular response to I/R include leukocyte–endothelial cell adhesion, transendothelial leukocyte migration, platelet–leukocyte aggregation, platelet-endothelial cell adhesion, and increased vascular permeability of albumin [1]. The I/R-induced increase in vascular permeability in venules is highly correlated with the number of adherent and emigrated leukocytes.

Leukocyte–Endothelial Responses in I/R Injury

The actions of the endothelial cell and the leukocytes are not independent from each other. A large amount of experimental data shows that xanthine oxidase-derived free radicals influence the interaction between endothelium and leukocytes, with the help of other chemoattractants such as platelet activating factor [37], leukotriene B_4 [38], and complement products [17]. The initial molecular interaction between leukocytes and the endothelium are transient and reversible, and is manifested as leukocyte rolling. The transition from rolling to firm adhesion of leukocytes requires their activation by soluble or surface-bound mediators. During this step, leukocytes respond to ligands on the endothelial cell surface by signaling, which in turn is rapidly followed by strong adhesion [1]. The apparent function of leukocyte activation is to elicit the adhesive function of integrin, which has been shown in a reconstituted in vitro system [39]. Once integrin function is activated, these molecules play a key role in mediating the strong adhesion of leukocytes to the endothelial surface. Firm adhesion of leukocytes (sticking) is largely CD18-dependent (β2 integrin) [40], and mediated by integrin–intracellular adhesion molecule (ICAM) pairs CD11a/CD18–ICAM-1, CD11a/CD18–ICAM-2, and CD11b/CD18–ICAM-1 [41,42]. Transendothelial migration is inhibited by anti-ICAM-1 monoclonal antibodies, suggesting that this endothelial adhesion molecule is required for migration of leukocytes beneath the endothelial cell layer [43].

The effects of I/R on leukocyte–endothelial cell adhesion have been examined in several animal models. Several methods have been used to define the role of leukocytes in I/R injury, including monitoring the enzyme myeloperoxidase (MPO), which occurs predominantly in neutrophils, appearance of radiolabeled neutrophils, and depletion of neutrophils. Complete arterial inflow occlusion has been shown to stimulate a 30- to 35-fold increase in adhering neutrophils in the microcirculation during the reperfusion period and smaller increments (2- to 10-fold) in neutrophil adhesion are observed during reperfusion period in models with incomplete ischemia [44,45]. Recent studies provide direct evidence for an increased leukocyte–endothelial interaction in the systemic microcirculation following I/R injury. Neutrophils entering recently reperfused tissue become activated, increase their synthesis of oxygen metabolites and proteolytic enzymes, and become more adhesive to the endothelium [1,2,10,13]. Neutrophils induce injury by the secretion of proteolytic enzymes such as elastase, probably in conjunction with neutrophil-generated oxidation products such as HOCl and H_2O_2. These result in lysis of essential structural matrix proteins, including collagen and fibronectin, leading to increased microvascular permeability [46].

Platelet–Endothelial Responses in I/R Injury

Although platelets lack a nucleus, they possess cellular machinery comparable to leukocytes in many aspects. Our group and other groups have shown that platelets also contribute to I/R injury by interacting with vascular endothelium [14,15,47]. Published evidence indicates that platelet adhesion is evident in the early [14] and late phases of reperfusion [15]. Platelets are equipped with several adhesion molecules that are required for cell–cell interaction, such as P-selectin, PECAM-1, and several integrins [13–15]. P-selectin, which is stored in platelets and endothelial cells, is rapidly expressed on the cell surface after I/R [14,15]. P-selectin, a key mediator of I/R-induced leukocyte adhesion, also appears to be important in mediating the interaction between platelets and activated endothelium in I/R [13,14,48]. P-selectin expression on endothelial cells and platelets can be elicited by a number of chemical mediators during I/R, such as oxygen radicals and cytokines [49,50]. These chemicals are also known to be released from activated endothelial cells, platelets, and leukocytes during I/R. Several other mediators such as thromboxane A_2, serotonin, adenosine diphosphate, platelet activating factor, thrombin, and tissue factor may also be involved either directly or indirectly with the platelet–endothelial cell adhesion induced by I/R [13,36]. Our recent study indicates that platelet, rather then endothelial, P-selectin is more important in I/R-induced platelet adhesion in the later stage of reperfusion [14].

Interactions between platelets and leukocytes have also frequently been observed in the postischemic microcirculation (Fig. 2) [14,47,48]. These interactions appear to be mediated mainly by platelet P-selectin and may result in reciprocal activation, resulting in an increased P-selectin expression by platelets [13,14,47,48] and increased β_2 integrin expression, and enhanced superoxide generation by leukocytes [13,14,48]. Such cell–cell interactions between leukocytes and platelets are likely to increase with prolonged reperfusion due to the increased expression of transcription-dependent mediators of inflammation. P-selectin glycoprotein ligand-1 (PSGL-1) is another possible receptor for platelet adhesion with either leukocyte or endothelium during I/R [13,14]. This view is supported by the expression of PSGL-1 on platelets [51] and reports of PSGL-1-mediated platelet adhesion [14]. P-selectin glycoprotein ligand-1 blocking antibody treatment in mice prior to I/R showed significantly decreased platelet adhesion to endothelium [14], indicating P-selectin–PSGL-1 interactions.

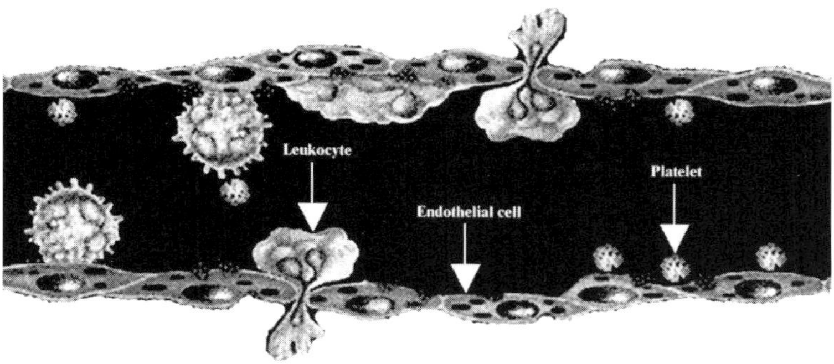

Fig. 2. Schematic illustration of the adhesion of leukocytes and platelets to venular endothelial cells

Other Factors Influencing the Microvascular Responses to I/R

Recent studies strongly suggest that risk factors such as hypercholesterolemia, hypertension, and diabetes enhance the vulnerability of the microvasculature to the injurious effect of I/R. This enhanced vulnerability to I/R-induced microvascular dysfunction is often manifested as an amplification of the inflammatory cell–cell interactions, diminished endothelial barrier function, and enhanced oxidant production [1,2]. Hypercholesterolemia alters endothelial cell function in all segments of the microvasculature. Arterial vessels exhibit an attenuated endothelium-dependent relaxation to vasodilator stimuli. An accelerated production of proinflammatory mediators by endothelial cells has been noted in hypercholesterolemic animals subjected to I/R injury [19]. Hypercholesterolemia also exaggerates I/R-induced capillary dysfunction and endothelial barrier dysfunction [31]. A larger number of rolling, adherent, and emigrating leukocytes as well as platelet–leukocyte aggregation have been observed following I/R in hypercholesterolemic animals [37]. Ischemia–reperfusion also elicits a more marked increase in albumin leakage and oxidant production in venules of hypercholesterolemic animals, and it has been proposed that a more pronounced diminution of NO bioavailability accounts for the exaggerated venular responses to I/R [52].

Chronic arterial hypertension either enhances or attenuates the responses elicited by inflammatory stimuli, such as I/R. Our group has compared the microvascular responses to I/R in spontaneously hypertensive (SHR) and normotensive rats [53]. We have found that the microvascular alterations normally elicited by I/R do not suffer significantly between SHR and normotensive rats. The number of firmly adherent and emigrated leukocytes, and platelet–leukocyte aggregation in postischemic venules of SHR, appear to be quite similar to those normotensive rats response. However, albumin leakage from postcapillary venules appears to be more profoundly enhanced by reperfusion in SHR than normotensive rats. It has been shown by our group that the magnitude of albumin leakage in venules elicited by I/R is directly proportional to the number of adherent and emigrated leukocytes [53]. In that regard, our findings in SHR demonstrate that, although the intensity of I/R-induced leukocyte recruitment in venules of SHR is no greater than that observed in normotensive rats, the endothelial barrier in venules of SHR appears to be more vulnerable to the leukocyte-mediated damage induced by I/R [53,54]. This enhanced vulnerability may due to number of factors observed in SHR, such as that neutrophils degranulate more readily as well as produce more superoxide [55], and may produce less NO as a consequence of reduced capacity to defend against neutrophil derived superoxide.

Diabetes mellitus has a broad range of deleterious effects on blood vessels and it appears to influence the responses of the microvasculature to I/R. Our group showed that in diabetic rats, there was a significant enhancement in the leukocyte rolling after ischemia, an effect that would facilitate adherent and emigrating leukocytes, more albumin leakage, and accelerated formation of oxygen radical formation [56]. Furthermore, diabetic animals showed pronounced pulmonary leak after sham operation alone and significant tissue injury following I/R [57]. Unlike hypertensive or hypercholesterolemic animals, albumin leak in diabetic animals do not result in a more intense albumin extravasation response to I/R. Glucose metabolism via polyol pathway, protein kinase C activation, and altered nitric oxide production by endothelial cells may have been involved as leukocyte-independent mechanisms to explain the increased microvascular permeability in diabetes.

In summary, reperfusion of ischemic tissue is associated with microvascular dysfunction that is manifested as impaired endothelium-dependent dilation in arterioles, enhanced fluid filtration and leukocyte plugging in capillaries, and the trafficking of leukocytes and plasma protein extravasation in postcapillary venules. Activated microvascular endothelial cells produce more oxygen radicals and less nitric oxide and this results in increased production proinflammatory mediators. Leukocyte–endothelial cell, leukocyte–platelet–endothelial cell, and platelet–endothelial cell interactions play a major role in the microvascular responses elicited by I/R. Some of the known risk factors for cardiovascular diseases such as hypercholesterolemia, hypertension, and diabetes appear to exaggerate many of the microvascular responses elicited by I/R.

References

1. Carden DL, Granger DN (2000) Pathophysiology of ischaemia-reperfusion injury. J Pathol 190:255–266
2. Granger DN (1999) Ischemia-reperfusion: mechanisms of microvascular dysfunction and the influence of risk factors for cardiovascular disease. Microcirculation 6:167–178
3. Zimmerman BJ, Granger DN (1994) Mechanisms of reperfusion injury. Am J Med Sci 307:284–292
4. Haglund U, Bulkley GB, Granger DN (1987) On the pathophysiology of intestinal ischemic injury. Clinical review. Acta Chir Scand 153:321–324
5. Parks DA, Bulkley GB, Granger DN (1983) Role of oxygen free radicals in shock, ischemia, and organ preservation. Surgery 94:428–432
6. Granger DN, Hollwarth ME, Parks DA (1986) Ischemia-reperfusion injury: role of oxygen-derived free radicals. Acta Physiol Scand Suppl 548:47–63
7. Grisham MB, Granger DN, Lefer DJ (1998) Modulation of leukocyte-endothelial interactions by reactive metabolites of oxygen and nitrogen: relevance to ischemic heart disease. Free Radic Biol Med 25:404–433

8. Granger DN, Kvietys PR, Perry MA (1993) Leukocyte-endothelial cell adhesion induced by ischemia and reperfusion. Can J Physiol Pharmacol 71:67–75
9. Bienvenu K, Granger DN (1993) Leukocyte adhesion in ischemia/reperfusion. Blood Cells 19:279–288
10. Cooper D, Stokes KY, Tailor A, et al (2002) Oxidative stress promotes blood cell-endothelial cell interactions in the microcirculation. Cardiovasc Toxicol 2:165–180
11. Parratt JR (1988) Pathophysiology of myocardial ischemia: importance of platelet-vessel wall interactions. Cardiovasc Drugs Ther 2:35–40
12. Mehta JL, Nicolini FA, Donnelly WH, et al (1992) Platelet-leukocyte-endothelial interactions in coronary artery disease. Am J Cardiol 69:8B–13B
13. Krieglstein CF, Granger DN (2001) Adhesion molecules and their role in vascular disease. Am J Hypertens 14:44S–54S
14. Cooper D, Chitman KD, Williams MC, et al (2003) Time-dependent platelet-vessel wall interactions induced by intestinal ischemia-reperfusion. Am J Physiol 284:G1027–G1033
15. Russell J, Cooper D, Tailor A, et al (2003) Low venular shear rates promote leukocyte-dependent recruitment of adherent platelets. Am J Physiol 284:G123–G129
16. Korthuis RJ, Anderson DC, Granger DN (1994) Role of neutrophil-endothelial cell adhesion in inflammatory disorders. J Crit Care 9:47–71
17. Anaya-Prado R, Toledo-Pereyra LH, Lentsch AB, et al (2002) Ischemia/reperfusion injury. J Surg Res 105:248–258
18. Banda MA, Lefer DJ, Granger DN (1997) Postischemic endothelium-dependent vascular reactivity is preserved in adhesion molecule-deficient mice. Am J Physiol 273:H2721–2725
19. Harrison DG (1997) Cellular and molecular mechanisms of endothelial cell dysfunction. J Clin Invest 100:2153–2157
20. Furchgott RF, Cherry PD, Zawadzki JV, et al (1984) Endothelial cells as mediators of vasodilation of arteries. J Cardiovasc Pharmacol 6:S336–S343
21. Inoue M, Sato EF, Park AM, et al (2000) Cross-talk between NO and oxyradicals, a supersystem that regulates energy metabolism and survival of animals. Free Radic Res 33:757–770
22. Korthuis RJ, Granger DN (1993) Reactive oxygen metabolites, neutrophils, and the pathogenesis of ischemic-tissue/reperfusion. Clin Cardiol 16:I19–126
23. Granger DN (1988) Role of xanthine oxidase and granulocytes in ischemiareperfusion injury. Am J Physiol 255:H1269–H1275
24. Grisham MB, Granger DN (1989) Metabolic sources of reactive oxygen metabolites during oxidant stress and ischemia with reperfusion. Clin Chest Med 10:71–81
25. Granger DN, Kubes P (1996) Nitric oxide as antiinflammatory agent. Methods Enzymol 269:434–442
26. Loughrey JP, Laffey JG, Moore BJ, et al (2003) Interleukin-1 beta rapidly inhibits aortic endothelium-dependent relaxation by a DNA transcription-dependent mechanism. Crit Care Med 31:910–915
27. Matsuki T, Duling BR (2000) TNF-alpha modulates arteriolar reactivity secondary to a change in intimal permeability. Microcirculation 7:411–418
28. Bertuglia S, Giusti A (2003) Microvascular oxygenation, oxidative stress, NO suppression and superoxide dismutase during postischemic reperfusion. Am J Physiol 285:H1064–H1071
29. Menger MD, Steiner D, Messmer K (1992) Microvascular ischemia-reperfusion injury in striated muscle: significance of "no reflow". Am J Physiol 263:H1892–H1900

30. Vollmar B, Menger MD (1997) Assessment of microvascular oxygen supply and tissue oxygenation in hepatic ischemia/reperfusion. Adv Exp Med Biol 428:403–408
31. Harris NR, Granger DN (1996) Neutrophil enhancement of reperfusion-induced capillary fluid filtration associated with hypercholesterolemia. Am J Physiol 271: H1755–H1761
32. Dayton C, Yamaguchi T, Warren A, et al (2002) Ischemic preconditioning prevents postischemic arteriolar, capillary, and postcapillary venular dysfunction: signaling pathways mediating the adaptive metamorphosis to a protected phenotype in preconditioned endothelium. Microcirculation 9:73–89
33. Jerome SN, Kong L, Korthuis RJ (1994) Microvascular dysfunction in postischemic skeletal muscle. J Invest Surg 7:3–16
34. Horie Y, Wolf R, Anderson DC, et al (1997) Hepatic leukostasis and hypoxic stress in adhesion molecule-deficient mice after gut ischemia/reperfusion. J Clin Invest 99: 781–788
35. Fitzal F, DeLano FA, Young C, et al (2002) Early capillary no-reflow during low-flow reperfusion after hind limb ischemia in the rat. Ann Plast Surg 49:170–180
36. Kubes P, Granger DN (1996) Leukocyte-endothelial cell interactions evoked by mast cells. Cardiovasc Res 32:699–708
37. Mori N, Horie Y, Gerritsen ME, et al (1999) Ischemia-reperfusion induced microvascular responses in LDL-receptor -/- mice. Am J Physiol 276:H1647–H1654
38. Bienvenu K, Russell J, Granger DN (1992) Leukotriene B4 mediates shear rate-dependent leukocyte adhesion in mesenteric venules. Circ Res 71:906–911
39. Lawrence MB, Springer TA (1991) Leukocytes roll on a selectin at physiologic flow rates: distinction from and prerequisite for adhesion through integrins. Cell 65: 859–873
40. Arfors KE, Lundberg C, Lindbom L, et al (1987) A monoclonal antibody to the membrane glycoprotein complex CD18 inhibits polymorphonuclear leukocyte accumulation and plasma leakage in vivo. Blood 69:338–340
41. Perry MA, Granger DN (1991) Role of CD11/CD18 in shear rate-dependent leukocyte-endothelial cell interactions in cat mesenteric venules. J Clin Invest 87:1798–1804
42. Ley K (2001) Pathways and bottlenecks in the web of inflammatory adhesion molecules and chemoattractants. Immunol Res 24:87–95
43. Kishimoto TK, Warnock RA, Jutila MA, et al (1991) Antibodies against human neutrophil LECAM-1 (LAM-1/leu-8/DREG-56 antigen) and endothelial cell ELAM-1 inhibit a common CD18-independent adhesion pathway in vitro. Blood 78:805–811
44. Erlansson M, Bergqvist D, Persson NH, et al (1991) Modification of postischemic increase of leukocyte adhesion and vascular permeability in the hamster by Iloprost. Prostaglandins 41:157–168
45. Granger DN, Benoit JN, Suzuki M, et al (1989) Leukocyte adherence to venular endothelium during ischemia-reperfusion. Am J Physiol 257:G683–G688
46. Welbourn CR, Goldman G, Paterson IS, et al (1991) Pathophysiology of ischaemia reperfusion injury: central role of the neutrophil. Br J Surg 78:651–655
47. Salter JW, Krieglstein CF, Issekutz AC, et al (2001) Platelets modulate ischemia/reperfusion-induced leukocyte recruitment in the mesenteric circulation. Am J Physiol 281:G1432–G1439
48. Massberg S, Enders G, Leiderer R, et al (1998) Platelet-endothelial cell interactions during ischemia/reperfusion: the role of P-selectin. Blood 92:507–515
49. Takano M, Meneshian A, Sheikh E, et al (2002) Rapid upregulation of endothelial P-selectin expression via reactive oxygen species generation. Am J Physiol 283:H2054–H2061

50. Horie Y, Chervenak RP, Wolf R, et al (1997) Lymphocytes mediate TNF-alpha-induced endothelial cell adhesion molecule expression: studies on SCID and RAG-1 mutant mice. J Immunol 159:5053–5062
51. Frenette PS, Denis CV, Weiss L, et al (2000) P-Selectin glycoprotein ligand 1 (PSGL-1) is expressed on platelets and can mediate platelet-endothelial interactions in vivo. J Exp Med 191:1413–1422
52. Kurose I, Wolf RE, Grisham MB, et al (1998) Hypercholesterolemia enhances oxidant production in mesenteric venules exposed to ischemia/reperfusion. Arterioscler Thromb Vasc Biol 18:1583–1588
53. Kurose I, Wolf R, Cerwinka W, et al (1999) Microvascular responses to ischemia/reperfusion in normotensive and hypertensive rats. Hypertension 34:212–216
54. Cerwinka WH, Granger DN (2001) Influence of hypercholesterolemia and hypertension on ischemia-reperfusion induced P-selectin expression. Atherosclerosis 154:337–344
55. Komatsu S, Panes J, Russell JM, et al (1997) Effects of chronic arterial hypertension on constitutive and induced intercellular adhesion molecule-1 expression in vivo. Hypertension 29:683–689
56. Panes J, Kurose I, Rodriguez-Vaca D, et al (1996) Diabetes exacerbates inflammatory responses to ischemia-reperfusion. Circulation 93:161–167
57. Lawson S, Ward DT, Conner C, et al (2002) Diabetic hyperglycemia: a facilitating factor in systemic capillary leak. J Surg Res 105:95–101

Analysis of Trigger Mechanisms for Inflammation in Cardiovascular Disease: Application to Shock and Multiorgan Failure

Geert W. Schmid-Schönbein[1] and Tony E. Hugli[2]

Summary. Cell activation in the microcirculation and inflammatory mediators have become a central focus of research into many cardiovascular diseases. However, the possible trigger mechanisms of inflammation remain less well understood. We summarize here a series of studies designed to investigate the origin of inflammation in acute physiological shock, a potentially lethal condition. A series of basic studies demonstrated that a major source of the inflammatory mediators in shock, produced in an ischemic intestine or after endotoxin administration, is due to the action of digestive enzymes. Many of these enzymes are proteases derived from the pancreas. The proteolytic and lipolytic action of pancreatic enzymes leads to the release of inflammatory mediators from the wall of the intestine. These mediators are transported via the intestinal microcirculation and the lymphatics into the central circulation, where they can initiate an inflammatory cascade with eventual multi-organ failure. These observations offer an opportunity for testing novel interventions against the lethal outcome of shock.

Key words. Inflammation, Pancreatic digestive enzyme, Mechanotransduction

Introduction

In the past few decades, a set of in vivo observations has led toward a new hypothesis for a possible sequence of events that may lead to cardiovascular disease. Initially shown in a variety of experimental and small scale clinical

[1] Department of Bioengineering, Microcirculation Laboratory, The Whitaker Institute of Biomedical Engineering, University of California, San Diego, La Jolla, CA 92093-0412, USA
[2] California Toxicology Research Institute, 1989 Palomar Oaks Way, Carlsbad, CA 92009, USA

studies (see summary in [1]), that many cardiovascular diseases in man are accompanied by cell activation. The process manifests itself at the level of the microcirculation in form of a reaction cascade collectively designated also as the *inflammation response*. The inflammation can be independent of apparent infections, and was demonstrated in all mammalian species tested to date. The list includes a range of important clinical conditions, such as myocardial ischemia [2,3], stroke [4], atherosclerosis [5], patients with chronic venous disease [6–9], diabetics [10], and even individuals with risk factors for lesion formation [11], to name just a few.

For example, in coronary heart disease and stroke, inflammation is now a recognized risk independent of cholesterol levels [12–14]. Markers of inflammation, such as leukocyte count, C-reactive protein levels in the plasma, fibrinogen levels, or interleukin-6 serve as useful and in some cases predictive clinical indicators [15–22]. Recent evidence suggests that patients with arterial hypertension, a condition that in the past has not been conclusively associated with inflammation, have significant signs of inflammation [23], although the inflammatory cascade may be modified by depressed endothelial membrane adhesion mechanisms [24,25]. Anti-inflammatory measures may serve to delay the onset of lesion formation in many different medical problems.

Thus, inflammation has taken on a central stage in research on cardiovascular disease. One of the conditions with an especially severe form of inflammation is physiological shock [26]. Current interventions against inflammation in shock have been of limit utility. We will focus therefore on shock. The severity of inflammation in shock poses a special challenge as an acute condition. Studies are therefore being designed to identify and characterize the possible origin of cell activation and inflammation.

Cell Activation and The Inflammatory Cascade

Signs of activation may be detected on virtually every cell type in the circulation. Since clinical screening usually relies on circulating cells, our analysis has frequently focused on leukocytes. Activated leukocytes may not be freely circulating and may be trapped in the capillary network [27]. Therefore, we also have used naïve cells from nonsymptomatic donors and applied plasma of patients to these naïve donor cells in conjunction with appropriate control plasma samples.

Cell activation can be detected by a variety of methods. Relatively early forms of cell activation can be detected on circulating leukocytes in form of actin polymerization (pseudopod or lammellipod formation), oxygen free radical formation, expression of adhesion molecules, or cytoplasmic degranulation.

Activation of cells in the circulation causes profound alteration of rheological and cytotoxic cell properties with impairment of the microcirculation and initiation of the inflammatory cascade. Key events of the inflammatory cascade are

- Oxygen free radical release
- Membrane bleb formation
- Impairment of endothelial nitric oxide generation
- Calcium release into the cell cytoplasm and ion exchange
- Expression of membrane adhesion molecules, prothrombotic protein expression
- Production and release of inflammatory mediators
- Matrix metalloproteinase activation
- Cleavage of the glycocalyx
- Rolling and adhesion of neutrophils to endothelium, transvascular migration
- Inhibition of the fluid shear response
- Redistribution of interendothelial membrane adhesion molecules
- Enhancement of endothelial permeability
- Mast cell degranulation
- Cell shape changes with pseudopod formation and actin polymerization
- Leukocyte entrapment in capillaries
- Erythrocyte aggregation with membrane peroxidation
- Shift of vascular smooth muscle contraction
- Leukocyte/platelet attachment and aggregation
- Platelet adhesion to endothelium
- Monocyte and T-lymphocyte infiltration
- Reduction of capillary perfusion
- Microvascular stasis
- Upregulation of proinflammatory genes
- Downregulation of anti-inflammatory genes
- Synthesis of acute phase proteins
- Apoptosis
- Organ dysfunction

This list is by no means comprehensive. There is also a list of events that are associated with resolution of the inflammatory reaction. In selected organs there are specific reactions associated with inflammation. As more events during inflammation are defined at the molecular level, additional indicators will become available. Many of the events are likely to occur more or less simultaneously. In general, inflammation tends to be accompanied by a reduction of normal cell function, irrespective of the particular function of an organ. Inflammation in nonseptic situations has many of the characteristics observed in bacterial infections, so the inflammatory reaction may primarily be part of a normal defense function. While this picture offers the opportunity for many new interventions against the deleterious effects of cell activation and suppression of the inflammation, there is a need for improved understanding of the mechanisms that lead to cell activation in the first place.

Mechanisms for Cell Activation in the Circulation

What mechanisms then may initiate the inflammatory cascade? It is convenient to set up a catalog of general classes of mechanisms.

- Many cases of inflammation may be associated with the presence of an actual inflammatory mediator. The list of such possible candidates includes bacterial/viral/fungal sources, cytokines, histamine, oxidized products, complement fragments, lipid membrane fragments (LTB_4, PAF), thrombin, and numerous others. This list is extraordinary long and beyond the present scope of discussion.
- Another possibility for inflammation may be due to *depletion* of anti-inflammatory mediators (nitric oxide, interleukin-10, glucocorticoids, serum albumin, etc.). This list is considerably shorter, suggesting that the anti-inflammatory measures in the circulation are limited [28–30].
- Juxtacrine activation, a process that may be mediated by cell membrane contact and mediators such as platelet activating factor and oxygen free radicals [31].
- Oxygen depletion or any significant gas pressure or temperature transitions tend to stimulate an inflammatory response [32].
- Physiological fluid shear stress is also an important regulator of inflammation [33–37].

The current evidence suggests that in shock the *first* mechanism may be dominating due to the presence of actual inflammatory mediators in plasma.

Which Organ May Generate Inflammatory Mediators?

In the following we describe a sequence of studies designed to explore the *origin* of inflammation in shock.

Our analysis started with the observation that high levels of cell activation can be detected in plasma during early phases of shock. In experimental forms of shock (e.g., hemorrhagic shock) the cell activation is already detectable within an hour of ischemia, indicating a nongenomic origin of early inflammatory mediators in the plasma [38]. This is in line with the fact that any mammalian species can be exposed to shock, so there is no specific gene or gene product that could be uniquely associated with the shock syndrome.

But what could be the source of the inflammatory mediators? While in the past specific mediators have been proposed, such as endotoxin, cytokines, and lipid derived mediators, no conclusive evidence has been advanced to support this hypothesis. We therefore embarked on a fundamental analysis designed to identify the source of the inflammatory mediators that are so readily detected in plasma during shock. Our approach was to determine which particular organ or organs could serve as a source of inflammatory mediators in the plasma. Comparison of a wide variety of individual organ homogenates yielded a surprising result. While the homogenates of many organs yield low levels of cell activation, the pancreas stands out as a unique organ among all we have tested [39]. This observation is in line with the earlier analysis by Lefer and Glenn [40]. Pancreatic homogenates generate a powerful inflammatory mediator effect when tested in vivo. Not only did the homogenate serve to activate leukocytes, but it also caused direct cell apoptosis in the mesentery microcirculation even in the absence of leukocytes. When administered into the circulation this homogenate produced rapid mortality in 100% of the animals tested [41]. These mediators trigger many of the inflammatory events listed above.

Furthermore, incubation of any organ homogenate with the selected pancreatic digestive enzyme chymotrypsin leads to generation of powerful inflammatory mediators, as potent as the one from the pancreas. Trypsin, elastase, and lipase also generate potent inflammatory mediators, e.g., from intestinal homogenates, while nucleases have yielded low levels of inflammatory mediators in this experimental design [42]. Thus the evidence suggests that digestive enzymes synthesized in the pancreas as part of normal digestion play a central role in the generation of inflammatory mediators in shock. The indications from these studies suggest that it is less a particular organ than the presence of digestive enzymes that control the generation of inflammatory mediators. Thus, there is a need to examine the intestine, an organ known to play a central role in shock for many centuries.

What Mechanisms May Prevent Self-Digestion by Pancreatic Enzymes?

Digestive enzymes are synthesized in the pancreas as part of normal physiology. The enzymes are initially stored in lipid granules and synthesized in a proenzyme (zymogen) form until released via a pancreatic duct through the duodenum into the small intestine. At that stage, digestive enzymes are fully activated and have the ability to digest most biological molecules. Self-digestion is prevented by compartmentalization of these aggressive enzymes. The current evidence indicates that the mucosal epithelium on the intestinal villi form the major barrier to contain the digestive enzymes in the lumen of the intestine. Conditions such as ischemia and oxygen depletion in the intestinal microcirculation or the presence of excessive bacterial mediators serves to open the mucosal barrier, increase its permeability, and permit entry of digestive enzymes into the submucosal space. At this point, the digestive enzymes have access to cell surface and extracellular matrix proteins. Thus it is a common finding in various shock models that entire villi are cleaved during intestinal ischemia. Loss of intestinal villi exposes the submucosa and the remainder of the wall of the intestine to self-digestion by digestive enzymes.

Inhibition of Pancreatic Enzymes in the Lumen of the Intestine During Shock

Inhibition of pancreatic serine proteases in the *lumen* of the intestine leads to attenuation of humoral activator production as well as many of the deleterious sequelae that accompany shock, such as inflammation, the reduction of the central blood pressure, and early indicators of multi-organ failure. The protection was demonstrated in shock induced by occlusion of the mesentery artery with different protease inhibitors, such as ANGD or FOY [43,44]. Both the level of inflammatory mediators in the plasma as well as infiltration of leukocytes into intestine and liver could be significantly reduced. Luminal administration of the protease inhibitor is more effective than administration into the circulation since the bulk of the proteases are located in the intestinal lumen as part of nutrition [41,45]. Inhibition of the intraintestinal digestive proteases blocks also inflammation in the microcirculation of peripheral organs [46], the *systemic inflammatory response syndrome.*

In shock produced by splanchnic artery occlusion, the inhibition provided by blockade of pancreatic enzymes in the lumen of the intestine could not be further improved by blockade of xanthine oxidase with allopurinol [47]. In septic shock produced by administration of endotoxin, a condition which in the past has been attributed exclusively to the action of endotoxin, one can also demonstrate a reduction of the inflammation if the pancreatic enzymes in the intestine are blocked [48]. Thus, even in this condition, endotoxin may initially cause a rise in permeability of the mucosal lining in the intestine. But at a later stage of shock a substantial amount of inflammatory mediators may be derived from the intestine by pancreatic proteases, and cause the sustained symptoms after the initial endotoxin challenge.

A New Class of Inflammatory Mediators

What biochemical factors may act as inflammatory mediators? We have seen that both hydrophilic and hydrophobic compounds are present among the inflammatory mediators derived from the pancreas [26]. Both the hydrophobic and hydrophilic components are lethal if administered into the circulation, but the hydrophilic fraction does not activate neutrophils [49]. There is a spectrum of mediators with molecular weight below 10 kDa [49] and in whole fractions of pancreatic homogenates below 5 kDa [39]. In lymphatic fluid, higher molecular weight compounds have also been detected [50]. The significance of these previously undescribed classes of biochemical mediators in other types of inflammatory reactions remains to be explored.

Conclusion

One of the important issues in current cardiovascular research is the analysis of the trigger mechanisms that serve to initiate the inflammatory cascade. Each disease may have its unique trigger mechanisms. We present here an analysis of shock, one of the most lethal conditions with severe forms of inflammation. The data suggest that the origin of the inflammatory markers may be derived from the lack of maintaining compartmentalization and thus permitting direct action of digestive enzymes in ischemic tissues.

Shock may be a process by which normal pancreatic enzymes, whether in the lumen of the intestine or elsewhere, are no longer properly contained and are free to attack tissue. The current evidence supports this hypothesis of an enzymatic mechanism. Shock and the release of inflammatory mediators into the portal veins and the central circulation may be the consequence of self-digestion by pancreatic and other cellular enzymes. The utility of intra-intestinal digestive enzyme inhibition under selected clinical conditions encountered in intensive care remains to be further explored and confirmed.

Acknowledgment. The research summarized in this chapter was supported by grants HL 67825 and HL 43026 from the U.S. NHLBI.

References

1. Schmid-Schönbein GW, Granger ND (eds) (2003) Molecular basis of microcirculatory disorders. Springer, Berlin Heidelberg New York London Paris
2. Chang RRK, Chien NTY, Chen C-H, et al (1992) Spontaneous activation of circulating granulocytes in patients with acute myocardial and cerebral diseases. Biorheology 29:549–561

3. Kassirer M, Zeltser D, Prochorov V, et al (1999) Increased expression of the CD11b/CD18 antigen on the surface of peripheral white blood cells in patients with ischemic heart disease: further evidence for smoldering inflammation in patients with atherosclerosis. Am Heart J 138:555–559
4. Grau AJ, Berger E, Sung K-LP, et al (1992) Granulocyte adhesion, deformability, and superoxide formation in acute stroke. Stroke 22:33–39
5. Ross R (1999) Atherosclerosis—an inflammatory disease. N Engl J Med 340:115–126
6. Takase S, Schmid-Schönbein G, Bergan JJ (1999) Leukocyte activation in patients with venous insufficiency. J Vasc Surg 30:148–156
7. Ono T, Bergan JJ, Schmid-Schönbein GW, et al (1998) Monocyte infiltration into venous valves. J Vasc Surg 27:158–166
8. Smith PD (2001) Update on chronic-venous-insufficiency-induced inflammatory processes. Angiology 52(suppl 1):S35–S42
9. Jünger M, Steins A, Hahn M, et al (2000) Microcirculatory dysfunction in chronic venous insufficiency (CVI). Microcirculation 7:S3–S12
10. Hokama JY, Ritter LS, Davis-Gorman G, et al (2000) Diabetes enhances leukocyte accumulation in the coronary microcirculation early in reperfusion following ischemia. J Diabetes Complications 14:96–107
11. Pitzer JE, Del Zoppo GJ, Schmid-Schönbein GW (1996) Neutrophil activation in smokers. Biorheology 33:45–58
12. Pearson TA, Mensah GA, Alexander RW, et al (2003) Markers of inflammation and cardiovascular disease: application to clinical and public health practice: A statement for healthcare professionals from the Centers for Disease Control and Prevention and the American Heart Association. Circulation 107:499–511
13. Schumacher A, Seljeflot I, Sommervoll L, et al (2002) Increased levels of markers of vascular inflammation in patients with coronary heart disease. Scand J Clin Lab Invest 62:59–68
14. del Zoppo G, Ginis I, Hallenbeck JM, et al (2000) Inflammation and stroke: putative role for cytokines, adhesion molecules and iNOS in brain response to ischemia. Brain Pathol 10:95–112
15. Danesh J, Whincup P, Walker M, et al (2000) Low grade inflammation and coronary heart disease: prospective study and updated meta-analyses. Br Med J 321:199–204
16. Lowe GD, Yarnell JW, Rumley A, et al (2001) C-reactive protein, fibrin D-dimer, and incident ischemic heart disease in the Speedwell study: are inflammation and fibrin turnover linked in pathogenesis? Arterioscler Thromb Vasc Biol 21:603–610
17. Hoffmeister A, Rothenbacher D, Bazner U, et al (2001) Role of novel markers of inflammation in patients with stable coronary heart disease. Am J Cardiol 87:262–266
18. Sinisalo J, Paronen J, Mattila KJ, et al (2000) Relation of inflammation to vascular function in patients with coronary heart disease. Atherosclerosis 149:403–411
19. Ridker PM, Hennekens CH, Buring JE, et al (2000) C-reactive protein and other markers of inflammation in the prediction of cardiovascular disease in women. N Engl J Med 342:836–843
20. Yudkin JS, Kumari M, Humphries SE, et al (2000) Inflammation, obesity, stress and coronary heart disease: is interleukin-6 the link? Atherosclerosis 148:209–214
21. Koenig W, Sund M, Frohlich M, et al (1999) C-Reactive protein, a sensitive marker of inflammation, predicts future risk of coronary heart disease in initially healthy middle-aged men: results from the MONICA (Monitoring Trends and Determinants in Cardiovascular Disease) Augsburg Cohort Study, 1984 to 1992. Circulation 99:237–242

22. Lindsberg PJ, Grau AJ (2003) Inflammation and infections as risk factors for ischemic stroke. Stroke 34:2518–2532
23. Lacy F, O'Connor DT, Schmid-Schönbein GW (1998) Plasma hydrogen peroxide production in hypertensives and normotensive subjects at genetic risk of hypertension. J Hypertens 16:291–303
24. Suzuki H, Schmid-Schönbein GW, Suematsu M, et al (1994) Impaired leukocyte-endothelial cell interaction in spontaneously hypertensive rats. Hypertension 24: 719–727
25. Suzuki H, Zweifach BW, Forrest MJ, et al (1995) Modification of leukocyte adhesion in spontaneously hypertensive rats by adrenal corticosteroids. J Leukocyte Biol 57:20–26
26. Schmid-Schönbein GW, Kistler EB, Hugli TE (2001) Mechanisms for cell activation and its consequences for biorheology and microcirculation: Multi-organ failure in shock. Biorheology 38:185–201
27. Shen K, Chavez-Chavez R, Loo AKL, et al (1995) Interpretation of leukocyte activation measurements from systemic blood vessels. In: Caplan LR (ed) Cerebrovascular diseases. 19th Princeton Stroke Conference, Moskowitz, MA, Chapter 6. Butterworth-Heinemann, Boston, pp 59–73
28. Kubes P, Suzuki M, Granger DN (1991) Nitric oxide: An endogenous modulation of leukocyte adhesion. Proc Natl Acad Sci USA 88:4651–6455
29. Froen JF, Munkeby BH, Stray-Pedersen B, et al (2002) Interleukin-10 reverses acute detrimental effects of endotoxin-induced inflammation on perinatal cerebral hypoxia-ischemia. Brain Res 942:87–94
30. Perretti M, Ahluwalia A (2000) The microcirculation and inflammation: site of action for glucocorticoids. Microcirculation 7:147–161
31. Zimmerman GA, McIntyre TM, Prescott SM (1996) Adhesion and signaling in vascular cell–cell interactions. J Clin Invest 98:1699–1702
32. Shah S, Allen J, Wood JG, et al (2003) Dissociation between skeletal muscle microvascular PO_2 and hypoxia-induced microvascular inflammation. J Appl Physiol 94: 2323–2329
33. Nagel T, Resnick N, Atkinson WJ, et al (1994) Shear stress selectively upregulates intercellular adhesion molecule-1 expression in cultured human vascular endothelial cells. J Clin Invest 94:885–891
34. Sheikh S, Rainger GE, Gale Z, et al (2003) Exposure to fluid shear stress modulates the ability of endothelial cells to recruit neutrophils in response to tumor necrosis factor-alpha: a basis for local variations in vascular sensitivity to inflammation. Blood 102:2828–2834
35. Chen BP, Li YS, Zhao Y, et al (2001) DNA microarray analysis of gene expression in endothelial cells in response to 24-h shear stress. Physiol Genomics 7:55–63
36. Fukuda S, Yasu T, Predescu DN, et al (2000) Mechanisms for regulation of fluid shear stress response in circulating leukocytes. Circ Res 86:E13–E18
37. Ji JY, Jing H, Diamond SL (2003) Shear stress causes nuclear localization of endothelial glucocorticoid receptor and expression from the GRE promoter. Circ Res 92:279–285
38. Barroso-Aranda J, Zweifach BW, Mathison JC, et al (1995) Neutrophil activation, tumor necrosis factor, and survival after endotoxic and hemorrhagic shock. J Cardiovasc Pharmacol 25(suppl 2):s23–s29
39. Kistler EB, Hugli T, Schmid-Schönbein GW (2000) The pancreas as a source of cardiovascular cell activating factors. Microcirculation 7:183–192
40. Lefer AM, Glenn TM (1971) Role of the pancreas in the pathogenesis of circulatory shock. Adv Exp Med Biol 23:311–335

41. Kistler EB, Lefer AM, Hugli TE, et al (2000) Plasma activation during splanchnic arterial occlusion shock. Shock 14:30–34
42. Waldo SW, Rosario HS, Penn AH, et al (2003) Pancreatic digestive enzymes are potent generators of mediators for leukocyte activation and mortality. Shock 20:138–143
43. Mitsuoka H, Kistler EB, Schmid-Schönbein GW (2000) Generation of in vivo activating factors in the ischemic intestine by pancreatic enzymes. Proc Natl Acad Sci USA 97:1772–1777
44. Mitsuoka H, Kistler EB, Schmid-Schönbein GW (2002) Protease inhibition in the intestinal lumen: attenuation of systemic inflammation and early indicators of multiple organ failure in shock. Shock 17:205–209
45. Deitch EA, Shi HP, Lu Q, et al (2003) Serine proteases are involved in the pathogenesis of trauma shock-induced gut and lung injury. Shock 19:452–456
46. Fitzal F, DeLano FA, Young C, et al (2002) Pancreatic protease inhibition during shock attenuates cell activation and peripheral inflammation. J Vasc Res 39:320–329
47. Mitsuoka H, Schmid-Schönbein GW (2000) Mechanisms for blockade of in vivo activator production in the ischemic intestine and multi-organ failure. Shock 14: 522–527
48. Fitzal F, Delano FA, Young C, et al (2003) Pancreatic enzymes sustain systemic inflammation after an initial endotoxin challenge. Surgery 134:446–456
49. Kramp WJ, Waldo S, Schmid-Schönbein GW, et al (2003) Characterization of two classes of pancreatic shock factors: functional differences exhibited by hydrophilic and hydrophobic shock factors. Shock 20:356–362
50. Adams CA Jr, Xu DZ, Lu Q, et al (2001) Factors larger than 100kd in post-hemorrhagic shock mesenteric lymph are toxic for endothelial cells. Surgery 129: 351–363

Selectin-Dependent Leukocyte Interactions with Vascular Surfaces

Rodger P. McEver

Summary. Binding of selectins to cell-surface ligands mediates rolling of leukocytes on activated platelets and endothelial cells and on other leukocytes. These multicellular adhesive interactions contribute to immune surveillance and to responses to tissue injury and infection. The leukocyte mucin, P-selectin glycoprotein ligand-1 (PSGL-1), mediates interactions with L-selectin on other leukocytes and with P-selectin on activated platelets and endothelial cells. Both molecular and cellular features affect selectin-dependent rolling under flow. One important cellular feature is the extrusion of long, thin membrane tethers, which reduce force on adhesive bonds and increase the probability of forming additional bonds along the tether. The kinetic and mechanical properties of selectin-ligand interactions also modulate rolling. At low force regimes, increasing wall shear stress prolongs the lifetimes of bonds between PSGL-1 and either P-selectin or L-selectin; these counter-intuitive interactions are called catch bonds. Further increases in wall shear stress shorten lifetimes; these interactions are called slip bonds. Stabilization of catch bonds may account for the shear threshold requirement for L-selectin-dependent rolling.

Key words. Selectins, Inflammation, Leukocyte, Cell adhesion, Endothelial cell

---

Cardiovascular Biology Research Program, Oklahoma Medical Research Foundation, and Department of Biochemistry and Molecular Biology, Oklahoma Center for Medical Glycobiology, University of Oklahoma Health Sciences Center, Oklahoma City, OK 73104, USA

Introduction

The selectins are type I membrane glycoproteins that mediate adhesion of leukocytes and platelets to vascular surfaces [1,2]. L-selectin is expressed on most leukocytes. E-selectin is expressed on cytokine-activated endothelial cells. P-selectin is rapidly redistributed from membranes of secretory granules to the surfaces of activated platelets and endothelial cells. Each selectin has a membrane-distal C-type lectin domain, followed by an epidermal growth factor (EGF)-like motif, a series of consensus repeats, a transmembrane domain, and a short cytoplasmic tail. P- and E-selectin bind primarily to ligands on leukocytes, and P-selectin also interacts with ligands on platelets and some endothelial cells. L-selectin binds to ligands on endothelial cells of high endothelial venules (HEV) in lymph nodes and on other leukocytes.

Selectin Ligands

Like most lectins, selectins bind to a range of glycoconjugates with varying affinities. Key challenges have been to identify preferred glycoconjugates with higher affinity or avidity for selectins, to elucidate the biosynthetic pathways for selectin ligands, and to determine which glycoconjugates actually mediate cell adhesion to selectins under physiological flow. All selectins bind with low affinity to glycans with terminal components that include α2,3-linked sialic acid and α1,3-linked fucose, typified by the sialyl Lewis x (sLe^x) determinant (NeuAcα2,3Galβ1,3[Fucα1,3]GlcNAcβ1-R). Crystal structures of sLe^x bound to the lectin domains of P- and E-selectin reveal a network of interactions between the fucose, a single Ca^{2+} ion, and several amino acids, including those that coordinate the Ca^{2+}; this explains the Ca^{2+}-dependent nature of binding to fucosylated glycans [3]. The sialic acid and the galactose also interact with the lectin domain. Targeted disruption of the gene encoding the α1,3-fucosyltransferase Fuc-TVII in mice significantly decreases selectin-mediated leukocyte trafficking, and disruption of the genes for both Fuc-TVII and Fuc-TIV eliminates these interactions [4,5]. These studies suggest that virtually all physiologically relevant selectin ligands require α1,3-linked fucose.

P- and L-selectin, but not E-selectin, also bind in a Ca^{2+}-independent manner to sulfated glycans such as heparin, fucoidan, and sulfoglucuronyl glycolipids, which lack sialic acid or fucose [6]. These interactions suggested that sulfation of fucosylated glycoconjugates might enhance their interactions with P- and L-selectin. Subsequent studies confirmed this notion and documented that sulfation may occur either on tyrosines or on the glycans of a glycoprotein. L-selectin binds to a series of mucins expressed by HEV of lymph nodes. These mucins have many core-2 *O*-glycans capped with sLe^x. The mucins are sulfated on the C6 position of galactose and *N*-acetylglucosamine (GlcNAc) residues on numerous core-2 or extended core-1 *O*-glycans [7,8]. Conceivably the sLe^x and sulfate components could be contributed by a single core-2 branch or a single core-1 extension. Alternatively, the sLe^x and the sulfate could be separately contributed by a core-2 branch and core-1 extension on a bi-antennary *O*-glycan or by two clustered *O*-glycans. The precise structural basis for how these various components cooperate to optimize binding to L-selectin is not known.

The selectin ligand with the most clearly demonstrated biological functions is P-selectin glycoprotein ligand-1 (PSGL-1), which is expressed on leukocytes [1,9]. PSGL-1 is a transmembrane, homodimeric mucin bearing multiple *O*-glycans on serines and threonines. Antibody blocking studies and genetic deletion of PSGL-1 demonstrate that PSGL-1 is the dominant ligand for P-selectin and L-selectin on leukocytes. Studies with synthetic glycosulfopeptides indicate that P-selectin binds in a stereospecific manner to the N-terminal region of human PSGL-1 through recognition of tyrosine sulfate residues, adjacent peptide determinants, and fucose, galactose, and sialic residues on a properly positioned core-2 *O*-glycan [10,11]. The glycosulfopeptide must present sLe^x on a short core-2 *O*-glycan. Human PSGL-1 has a small number of these short, fucosylated *O*-glycans, but has more *O*-glycans bearing sLe^x on an extended, polyfucosylated core-2 branch [12]. However, a glycosulfopeptide with this extended glycan binds poorly to P-selectin [13]. Thus, optimal binding of P-selectin is critically dependent on the relative orientations of sLe^x and the sulfated peptide. The crystal structure of P-selectin complexed with a PSGL-1-derived glycosulfopeptide with sLe^x on a short core-2 *O*-glycan reveals a broad, shallow binding interface [3]. The Ca^{2+}-dependent interactions with sLe^x on the core-2 *O*-glycan are supplemented by Ca^{2+}-independent contacts with tyrosine sulfate and other amino acids. These additional contacts explain why P-selectin binds with much higher affinity to PSGL-1 than to sLe^x alone. Targeted deletion of the murine gene encoding Core2GlcNAcT-I, the major core-2 β1,6-*N*-acetylglucosaminyltransferase in leukocytes, eliminates binding of leukocytes to P-selectin [14,15]. This suggests that Core2GlcNAcT-I plays a key role in constructing the relevant core-2 *O*-glycan on PSGL-1.

Like human PSGL-1, murine PSGL-1 is a homodimeric sialomucin [16,17]. Monoclonal antibodies (mAbs) to N-terminal peptide epitopes of PSGL-1 block rolling of murine leukocytes on murine P-selectin in vitro and in vivo [18–20]. This suggests that P-selectin probably also binds to the N terminus of murine PSGL-1. However, the N terminus of murine PSGL-1 has a very different amino acid sequence than human PSGL-1 [17]. Murine PSGL-1 has potential sites for sulfation at Tyr-13 and -15 and for *O*-glycosylation at Thr-

14 and -17. Site-directed mutagenesis studies suggest that murine PSGL-1 requires sulfation of Tyr-13 and *O*-glycosylation of Thr-17 to bind optimally to P-selectin [21]. Because it uses only one tyrosine, murine PSGL-1 may rely more on other peptide components and *O*-glycosylation to bind to P-selectin than does human PSGL-1.

Enzymatic desialylation of murine leukocytes eliminates binding to E- and P-selectin [22], and selectins do not bind to leukocytes from mice that are deficient in Fuc-TVII and Fuc-TIV, the α1,3-fucosyltransferases normally expressed in these cells [23,24]. These combined data suggest that selectins recognize sLe^x-related glycans on murine leukocytes. However, many mAbs to sLe^x and Le^x fail to bind to murine leukocytes [25,26]. It has been widely assumed that unknown glycan modifications, perhaps unique to murine tissues, mask the epitopes for these mAbs. This assumption has not been critically tested, since direct structural characterization of glycans on murine leukocytes has not been performed. Furthermore, treatment of murine monocytic WEHI-3 cells with glycosidases or chlorate demonstrated that sialic acid modifications, α1,3-galactosylation, or sulfation do not mask epitopes for mAbs to sLe^x or Le^x [27]. WEHI-3 cells and murine neutrophils express low α1,3-fucosyltransferase activities compared to human promyelocytic HL-60 cells. Consistent with very low endogenous fucosylation, forced fucosylation of intact WEHI-3 cells or murine neutrophils by exogenous α1,3-fucosyltransferase Fuc-TVI and GDP-fucose creates many new epitopes for anti-sLe^x mAbs such as HECA-452 and CSLEX-1. Nevertheless, forced fucosylation of intact cells does not significantly augment their ability to bind to fluid-phase P- or E-selectin or to roll on immobilized P- or E-selectin under flow [27]. These data suggest that murine myeloid leukocytes fucosylate only a few specific glycans, which interact preferentially with P- and E-selectin. Even on human HL-60 cells, which express abundant sLe^x determinants, only a small minority of *O*-glycans on PSGL-1 are fucosylated, and perhaps only one or two of these *O*-glycans has the short core-2 *O*-glycan capped with sLe^x that cooperates with sulfated tyrosines and peptide components to bind optimally to P-selectin [12,13]. Therefore, both human and murine leukocytes may employ limited but specific α1,3-fucosylation to synthesize glycoconjugates that bind to selectins.

Identifying physiologically relevant glycoprotein ligands for E-selectin has been particularly difficult, because most cells roll on E-selectin if they are transfected with an expression vector encoding an α1,3-fucosyltransferase that introduces sLe^x epitopes on the cell surface. Thus, a frequent consideration is that E-selectin interacts indiscriminately with sLe^x-terminated glycans on many glycoproteins or glycolipids on leukocytes. However, genetic deletion of PSGL-1 in mice impairs leukocyte tethering to E-selectin in vitro and in vivo [28]. The residual leukocytes that tether roll equivalently to wild-type leukocytes. This demonstrates a novel function of PSGL-1 in tethering free-flowing leukocytes to E-selectin, but not in stabilizing subsequent rolling. A similar phenotype is seen in Core2GlcNAcT-I-deficient mice, suggesting that the sLe^x-capped core-2 *O*-glycans that interact with E-selectin are primarily on PSGL-1 [15]. These data establish PSGL-1 as a physiologically relevant glycoprotein ligand for all three selectins.

Regulation of Cell Rolling Under Flow

Rolling cell adhesion in the vasculature requires the rapid formation and breakage of adhesive bonds that are subjected to applied force [1]. Surface plasmon resonance measurements indicate that the association and dissociation kinetics of unstressed selectin-ligand bonds are rapid, although they vary considerably depending on the selectin and the ligand [29–31]. The lifetimes of transient leukocyte tethers on low-density selectins shorten in response to increasing wall shear stress. This is a characteristic of "slip bonds," where tensile force accelerates dissociation. However, low tensile forces prolong lifetimes of P-selectin or L-selectin bonds with PSGL-1, as measured by both atomic force microscopy and transient tether lifetimes [32,33]. This counterintuitive behavior is a characteristic of "catch bonds," which have not been demonstrated previously. Transitions between catch and slip bonds may help explain the requirement for a minimum wall shear stress below which cells do not roll on selectins. Bonds experience increasing forces as they move from the leading edge to the trailing edge of the rolling cell, potentially shifting from the catch bond regime to the slip bond regime. Catch bonds may act to prevent premature dissociation before the bonds arrive at the trailing edge, thereby maintaining stable rolling.

Leukocytes roll on selectins at nearly constant velocities over a wide range of wall shear stresses [34]. This "automatic braking system" has been ascribed to intrinsic molecular features of selectins and their ligands; higher wall shear stresses are postulated to overcome repulsive forces and increase bond formation. However, microspheres coupled with selectin ligands do not roll stably on selectins, whereas the same ligands coupled to cell surfaces confer stable rolling over a wide range of wall shear stresses [35]. Fixation of the cells before ligand coupling destabilizes rolling, as observed for ligand-coupled microspheres. This indicates that stable selectin-dependent rolling requires cellular features as well as the intrinsic molecular components of selectins and their ligands. These features may include cellular deformation, which increases the adhesive contact area [36], and extrusion of long membrane tethers, which reduces the force on tethers and allows the cell to slip or roll downstream from the tether [37].

The organization and orientation of selectins and their ligands on cell surfaces play major roles in modulating adhesion under flow [1]. Dimerization of P-selectin and PSGL-1 has little effect on the initial tethering of cells but does stabilize subsequent rolling [38]. If a cell is tethered by interactions of a dimeric selectin with a dimeric ligand, one bond may dissociate, leaving the cell still tethered. The first bond may then form again, stabilizing the tether and prolonging its lifetime. Replacement of the EGF domain of L-selectin with that of P-selectin alters the orientation of the lectin domain of L-selectin, allowing it to bind ligands more rapidly under flow. The result is more effective tethering and more stable rolling [39].

Clustering of selectins or selectin ligands through interactions with cytoskeletal proteins provides an additional mechanism to increase bond number, thereby reducing the force on individual bonds and prolonging tether lifetimes [1]. The cytoplasmic domain of L-selectin contains a membrane-distal binding site for α-actinin [40] and a membrane-proximal binding site for moesin [41]. Truncation of the region that includes the α-actinin binding site destabilizes rolling by shortening the lifetimes of adhesive tethers under flow [42]. Deletion of both the moesin and α-actinin binding sites virtually eliminates rolling, suggesting that both adaptors cooperate to link L-selectin to a stable network of actin filaments. L-selectin lacking the α-actinin binding site but retaining the moesin binding site is concentrated in microvilli like the wild-type protein [43]. Together, these data suggest that moesin bridges L-selectin and actin filaments in microvilli. The cytoplasmic domain of PSGL-1, which is also concentrated in microvilli, contains a membrane-proximal binding site for moesin [44]. Truncation of the cytoplasmic domain eliminates rolling of some PSGL-1-expressing cells on P-selectin [45]. It is possible that the cytoplasmic domain of PSGL-1 binds to more than one cytoskeletal adaptor, as does L-selectin. Further study is required to determine the contributions of specific portions of the cytoplasmic domains of L-selectin and PSGL-1 to their cell-surface organizations and adhesive functions.

References

1. McEver RP (2001) Adhesive interactions of leukocytes, platelets, and the vessel wall during hemostasis and inflammation. Thromb Haemost 86:746–756
2. Vestweber D, Blanks JE (1999) Mechanisms that regulate the function of the selectins and their ligands. Physiol Rev 79:181–213
3. Somers WS, Tang J, Shaw GD, Camphausen RT (2000) Insights into the molecular basis of leukocyte tethering and rolling revealed by structures of P- and E-selectin bound to SLe(X) and PSGL-1. Cell 103:467–479
4. Homeister JW, Thall AD, Petryniak B, et al (2001) The a(1,3)fucosyltransferases FucT-IV and FucT-VII exert collaborative control over selectin-dependent leukocyte recruitment and lymphocyte homing. Immunity 15:115–126

5. Smithson G, Rogers CE, Smith PL, et al (2001) Fuc-TVII is required for T helper 1 and T cytotoxic 1 lymphocyte selectin ligand expression and recruitment in inflammation, and together with Fuc-TIV regulates naive T cell trafficking to lymph nodes. J Exp Med 194:601–614
6. Varki A (1997) Selectin ligands: will the real ones please stand up? J Clin Invest 99:158–162
7. Hemmerich S, Bistrup A, Singer MS, et al (2001) Sulfation of L-selectin ligands by an HEV-restricted sulfotransferase regulates lymphocyte homing to lymph nodes. Immunity 15:237–247
8. Yeh JC, Hiraoka N, Petryniak B, et al (2001) Novel sulfated lymphocyte homing receptors and their control by a Corel extension beta 1,3-N-acetylglucosaminyltransferae. Cell 105:957–969
9. McEver RP, Cummings RD (1997) Role of PSGL-1 binding to selectins in leukocyte recruitment. J Clin Invest 100:485–492
10. Leppanen A, Mehta P, Ouyang YB, et al (1999) A novel glycosulfopeptide binds to P-selectin and inhibits leukocyte adhesion to P-selectin. J Biol Chem 274:24838–24848
11. Leppänen A, White SP, Helin J, et al (2000) Binding of glycosulfopeptides to P-selectin requires stereospecific contributions of individual tyrosine sulfate and sugar residues. J Biol Chem 275:39569–39578
12. Wilkins PP, McEver RP, Cummings RD (1996) Structures of the O-glycans on P-selectin glycoprotein ligand-1 from HL-60 cells. J Biol Chem 271:18732–18742
13. Leppänen A, Penttilä L, Renkonen O, et al (2002) Glycosulfopeptides with O-glycans containing sialylated and polyfucosylated polylactosamine bind with low affinity to P-selectin. J Biol Chem 277:39749–39759
14. Ellies LG, Tsuboi S, Petryniak B, et al (1998) Core 2 oligosaccharide biosynthesis distinguishes between selectin ligands essential for leukocyte homing and inflammation. Immunity 9:881–890
15. Sperandio M, Thatte A, Foy D, et al (2001) Severe impairment of leukocyte rolling in venules of core 2 glucosaminyltransferase-deficient mice. Blood 97:3812–3819
16. Lenter M, Levinovitz A, Isenmann S, et al (1994) Monospecific and common glycoprotein ligands for E- and P-selectin on myeloid cells. J Cell Biol 125:471–481
17. Yang J, Galipeau J, Kozak CA, et al (1996) Mouse P-selectin glycoprotein ligand-1: Molecular cloning, chromosomal localization, and expression of a functional P-selectin receptor. Blood 87:4176–4186
18. Borges E, Tietz W, Steegmaier M, et al (1997) P-selectin glycoprotein ligand-1 (PSGL-1) on T helper 1 but not on T helper 2 cells binds to P-selectin and supports migration into inflamed skin. J Exp Med 185:573–578
19. Borges E, Eytner R, Moll T, et al (1997) The P-selectin glycoprotein ligand-1 is important for recruitment of neutrophils into inflamed mouse peritoneum. Blood 90:1934–1942
20. Steegmaier M, Blanks JE, Borges E, et al (1997) P-selectin glycoprotein ligand-1 mediates rolling of mouse bone marrow-derived mast cells on P-selectin but not efficiently on E-selectin. Eur J Immunol 27:1339–1345
21. Xia L, Ramachandran V, McDaniel JM, et al (2003) N-terminal residues in murine P-selectin glycoprotein ligand-1 required for binding to murine P-selectin. Blood 101:552–559
22. Varki A (1994) Selectin ligands. Proc Natl Acad Sci USA 91:7390–7397
23. Maly P, Thall AD, Petryniak B, et al (1996) The a(1,3)Fucosyltransferase Fuc-TVII controls leukocyte trafficking through an essential role in L-, E-, and P-selectin ligand biosynthesis. Cell 86:643–653

24. Weninger W, Ulfman LH, Cheng G, et al (2000) Specialized contributions by alpha(1,3)-fucosyltransferase-IV and FucT-VII during leukocyte rolling in dermal microvessels. Immunity 12:665–676
25. Ito K, Handa K, Hakomori S (1994) Species-specific expression of sialosyl-Le(x) on polymorphonuclear leukocytes (PMN), in relation to selectin-dependent PMN responses. Glycoconj J 11:232–237
26. Thorpe SJ, Feizi T (1984) Species differences in the expression of carbohydrate differentiation antigens on mammalian blood cells revealed by immunofluorescence with monoclonal antibodies. Biosci Rep 4:673–685
27. Kobzdej MMA, Leppänen A, Ramachandran V, et al (2002) Discordant expression of selectin ligands and sialyl Lewis x-related epitopes on murine myeloid cells. Blood 100:485–494
28. Xia L, Sperandio M, Yago T, et al (2002) P-selectin glycoprotein ligand-1-deficient mice have impaired leukocyte tethering to E-selectin under flow. J Clin Invest 109:939–950
29. Mehta P, Cummings RD, McEver RP (1998) Affinity and kinetic analysis of P-selectin binding to P-selectin glycoprotein ligand-1. J Biol Chem 273:32506–32513
30. Nicholson MW, Barclay AN, Singer MS, et al (1998) Affinity and kinetic analysis of L-selectin (CD62L) binding to glycosylation-dependent cell-adhesion molecule-1. J Biol Chem 273:763–770
31. Wild MK, Huang MC, Schulze-Horsel U, et al (2001) Affinity, kinetics, and thermodynamics of E-selectin binding to E-selectin ligand-1. J Biol Chem 276:31602–31612
32. Marshall BT, Long M, Piper JW, et al (2003) Direct observation of catch bonds involving cell-adhesion molecules. Nature 423:190–193
33. Sarangapani KK, Yago T, Klopocki AG, et al (2004) Low force decelerates L-selectin dissociation from P-selectin glycoprotein ligand-1 and endoglycan. J Biol Chem: in press
34. Chen SQ, Springer TA (1999) An automatic braking system that stabilizes leukocyte rolling by an increase in selectin bond number with shear. J Cell Biol 144:185–200
35. Yago T, Leppänen A, Qiu H, et al (2002) Distinct molecular and cellular contributions to stabilizing selectin-mediated rolling under flow. J Cell Biol 158:787–799
36. Lei X, Lawrence MB, Dong C (1999) Influence of cell deformation on leukocyte rolling adhesion in shear flow. J Biomech Eng 121:636–643
37. Schmidtke DW, Diamond SL (2000) Direct observation of membrane tethers formed during neutrophil attachment to platelets or P-selectin under physiological flow. J Cell Biol 149:719–729
38. Ramachandran V, Yago T, Epperson TK, et al (2001) Dimerization of a selectin and its ligand stabilizes cell rolling and enhances tether strength in shear flow. Proc Natl Acad Sci USA 98:10166–10171
39. Dwir O, Kansas GS, Alon R (2000) An activated L-selectin mutant with conserved equilibrium binding properties but enhanced ligand recognition under shear flow. J Biol Chem 275:18682–18691
40. Pavalko FM, Walker DM, Graham L, et al (1995) The cytoplasmic domain of L-selectin interacts with cytoskeletal proteins via a-actinin: Receptor positioning in microvilli does not require interaction with a-actinin. J Cell Biol 129:1155–1164
41. Ivetic A, Deka J, Ridley A, et al (2002) The cytoplasmic tail of L-selectin interacts with members of the ezrin-radixin-moesin (ERM) family of proteins. J Biol Chem 277:2321–2329
42. Dwir O, Kansas GS, Alon R (2001) Cytoplasmic anchorage of L-selectin controls leukocyte capture and rolling by increasing the mechanical stability of the selectin tether. J Cell Biol 155:145–156

43. Kansas GS, Ley K, Munro JM, et al (1993) Regulation of leukocyte rolling and adhesion to high endothelial venules through the cytoplasmic domain of L-selectin. J Exp Med 177:833–838
44. Serrador JM, Urzainqui A, Alonso-Lebrero JL, et al (2002) A juxta-membrane amino acid sequence of P-selectin glycoprotein ligand-1 is involved in moesin binding and ezrin/radixin/moesin-directed targeting at the trailing edge of migrating lymphocytes. Eur J Immunol 32:1560–1566
45. Snapp KR, Heitzig CE, Kansas GS (2002) Attachment of the PSGL-1 cytoplasmic domain to the actin cytoskeleton is essential for leukocyte rolling on P-selectin. Blood 99:4494–4502

Lymphocyte Migration to the Intestinal Mucosa and its Relation to Mucosal Defense

Soichiro Miura[1], Yoshikazu Tsuzuki[1], Ryota Hokari[1], Hiroshi Nagata[2], and Hiromasa Ishii[2]

Summary. Lymphocyte recirculation is a key phenomenon in immunology. However, in vivo regulation of the homing phenomenon of lymphoid cells to the intestinal mucosa, and its pathophysiological role in intestinal inflammation, has not been clearly understood. In this chapter we summarize the dynamic process of lymphocyte–endothelium recognition in the lymphoid and nonlymphoid area of intestine under an intravital microscope using fluorescence-labeled lymphocytes and discuss how regulation of lymphocyte homing is disturbed or altered under inflammatory conditions. Microvessels of intestinal lamina propria efficiently attracted gut-derived T cells via the $\alpha 4\beta 7$/MAdCAM-1 system, but under tumor necrosis factor-α-induced inflamed conditions vascular cell adhesion molecule (VCAM)-1 was also significantly involved. Under physiological conditions there was little lymphocyte adherence to the colonic mucosa, but in inflamed colonic mucosa T-cell migration became significant, comparable to that in the small intestinal mucosa. The chemokine CCL25/TECK may play an important role in T-cell migration to uninflamed as well as inflamed small intestine, but not colon. In an animal model of chronic colitis enhanced upregulation of mucosal addressin cell adhesion molecule (MAdCAM)-1 levels in the colonic mucosa was demonstrated, and administration of anti-MAdCAM-1 antibody significantly attenuated the colonic injury, suggesting this adhesion molecule as a useful target for inflammatory bowel diseases.

Key words. Lymphocyte migration, Lamina propria lymphocyte, MAdCAM-1, Inflammatory bowel disease, TECK/CCL25

---

[1] Second Department of Medicine, National Defense Medical College, 3-2 Namiki, Tokorozawa, Saitama 359-8513, Japan

[2] Department of Internal Medicine, Keio University School of Medicine, 35 Shinanomachi, Shinjuku-ku, Tokyo 160-8582, Japan

Lymphocyte Migration in Peyer's Patches

Naïve lymphocytes generally can migrate very efficiently from the blood into secondary lymphoid tissues, such as Peyer's patches, by extravasating through high endothelial venules (HEVs). Bargatze et al. [1] have demonstrated that efficient arrest of naïve cells to Peyer's patch HEVs requires the sequential engagement of L-selectin to initial contact, $\alpha4\beta7$ to slow rolling, and lymphocyte function-associated antigen (LFA)-1 in conjunction with $\alpha4\beta7$/MAdCAM-1 to mediate activation-dependent arrest. Our previous data on lymphocyte behavior of nonactivated splenic T cells in Peyer's patches showed complete block of interaction of splenic lymphocytes with Peyer's patch HEVs by antibody against either L-selectin, α4-integrin, β7-integrin, or mucosal addressin cell adhesion molecule (MAdCAM)-1 is strong evidence for a multimolecular adhesion cascade in these HEVs [2]. The differential expression of adhesion-triggering chemokines, including CCL21, will control lymphocyte subset trafficking in Peyer's patches. In our results on rat Peyer's patches, B-cell arrest occurs in a wider range of vessels than T cells, including smaller venules observed inside follicles, whereas T cells preferentially accumulate in interfollicular HEVs [3]. It is known that T cells preferentially arrest in segments displaying high levels of luminal CCL21 in PP-HEVs.

After adhesion, the migration of lymphocytes from inside to outside of the venules (transendothelial migration) is observed, and eventually these cells migrate into the interstitium of Peyer's patches apart from the venular walls. Subsequently, T cells, especially $CD4^+$ cells, preferentially appear in parafollicular microlymphatics. This sequential migration process of naïve T cells in Peyer's patches could be modulated not only by chemokines, but also by various neurohumoral factors, including vasoactive intestinal peptide, nitric oxide, and ingested long-chain fatty acids [4–6].

Lymphocyte Migration in the Intestinal Mucosa

The intestinal lamina propria is the abundant source of lymphocytes involved in immune effector functions. However, there have been few reports of an in vivo observation study on how lymphocytes from intestinal mucosa migrate into the intestinal lamina propria itself (Fig. 1). The intravital microscopic procedure for monitoring the dynamic process of lymphocyte migration was used to investigate the possible contribution of cell adhesion molecules to the lymphocyte–endothelial cell adhesive interactions in the intestinal mucosa. Lamina proprial vessels of intestine are known to express MAdCAM-1, yet are thought to lack L-selectin-binding carbohydrates. Lamina propria lymphocytes (LPLs) were shown to express strongly CD69, an activation marker, β7,

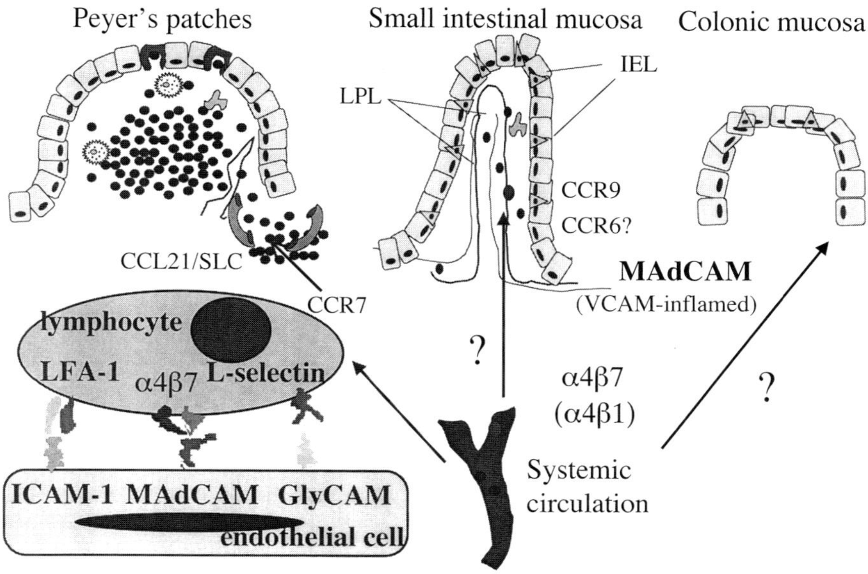

FIG. 1. Lymphocyte homing in Peyer's patches vs intestinal mucosa. In Peyer's patches three combinations of adhesion molecules (L-selectin–sLe^x, α4β7-integrin–MAdCAM-1, and αLβ2-integrin–ICAM-1) are responsible for T-cell migration with help of CC chemokine receptor 7. On the other hand, little information has been available on how lymphocyte–endothelial cell interaction occurs in the intestinal mucosal area. Our intravital observation showed that in both small intestinal or colonic mucosa α4β7-integrin–MAdCAM-1 interaction is crucial for T-lymphocyte migration in uninflamed as well as inflamed conditions. *MAdCAM*, mucosal addressin cell adhesion molecule; *LPL*, lamina propria lymphocytes; *IEL*, intestinal intraepithelial lymphocytes; *VCAM*, vascular cell adhesion molecule; *ICAM*, intracellular adhesion molecule; *LFA-1*, lymphocyte function-associated antigen-1; *GlyCAM*, glycosylation-dependent cell adhesion molecule

and a combinatorial epitope of α4β7, whereas there was little expression of L-selectin. Lamina propria lymphocytes accumulated abundantly in the microvessels of villus tips but not in the submucosal venules or postcapillary venules of Peyer's patches [2]. This may not be due to mechanical trapping in the capillaries but due to G-protein-mediated intracellular pathways, because the binding of LPLs to the lamina propria vessels was inhibited by the pretreatment with pertussis toxin. The accumulation of LPLs in the villus tip was almost completely blocked by anti-β7-integrin and was significantly inhibited by anti-MAdCAM-1 and anti-α4-integrin, but not by anti-CD11a or anti-L-selectin antibody [2]. The direct binding of α4β7 to MAdCAM-1 is thought to help LPLs migrate extremely efficiently to the intestinal lamina propria at least under physiological conditions. A critical role of β7-integrin in lymphocyte recruitment for formation of the gut-associated lymphoid tissue has been proposed. Wagner et al. [7] have reported that the number of $CD4^+$ cells was significantly reduced in the lamina propria of mice deficient for the β7-integrin. However, although the expression and function of β7 are major determinants of adhesion to MAdCAM-1 in lamina propria, the level of β7 expression may not exclusively account for lymphocyte preference to MAdCAM-1 or MAdCAM-1 expressed microvessels.

Intestinal intraepithelial lymphocytes (IELs) are a large cell population of cells in the epithelial layer of the small intestine. Although these cells have diverse origins, the normal IEL in the blood probably gain access to the intestinal mucosa, because injected peripheral T cells can fill the epithelial compartments and acquire the characteristics of IEL in adequate conditions. AlphaEβ7-integrin is expressed on 90% of the IEL and αEβ7 may mediate the binding of IEL to epithelial cells; however, how they reach the microvessels of the intestinal villus microcirculation is not known. Using isolated IELs from murine intestine, we demonstrated that IEL time-dependently accumulated in villus microvessels of the small intestine, but not to the postcapillary venules of Peyer's patches [8]. The accumulation of IEL in villus arcade was significantly, but partially inhibited by anti-β7-integrin or anti-MAdCAM-1 antibody, suggesting the possible role of other adhesion mechanisms as well.

Differences Between Small and Large Intestine, and Role of Chemokine

Relatively little is known about how recirculation of lymphocytes through the inflamed intestinal mucosa is regulated. It is becoming increasingly apparent that inflammation is associated with enhanced expression of endothelial cell adhesion molecules. However, relatively little is known how the different endothelial cell adhesion molecules contribute to T-lymphocyte recruitment in the intestinal mucosa under inflammatory conditions. Recent studies indicate that immune regulation in the colonic mucosa is different from that in the small intestine. For example, colonic IELs and LPLs appear to be less activated than their small intestinal counterparts. The difference may be related to the luminal environment of the bowel at these anatomic sites. However, there is relatively little quantitative information related to lymphocyte trafficking, especially for the colonic mucosa. Our recent study was conducted to investigate the dynamic process of T lymphocyte–endothelial cell adhesion in murine colonic mucosa and to examine how intestinal inflammation affects on lymphocyte interactions with microvessels [9]. In tumor necrosis factor (TNF)-α-stimulated colonic venules, and enhanced adhesion of LPLs was demonstrated. This increase was also significantly observed in the small intestinal mucosa, but the magnitude of the increased LPL adhesion was more significant in the colon (2.23 times) than in the small intestine (1.73 times). Immunohistochemistry revealed that after TNF-α-treatment, both MAdCAM-1 and vascular cell adhesion molecule (VCAM)-1 expression were increased in small intestinal and colonic mucosa, but there was a dominant increase in MAdCAM-1 in the colonic mucosa, that is consistent with the increased LPL adhesion to the inflamed colon. There was a significant MAdCAM-1 expression in the lamina propria and VCAM-1 expression in the submucosa of TNF-α-treated colon. The T-lymphocyte interaction in the colonic mucosa was significantly reduced by blocking monoclonal antibodies against either MAdCAM-1, α4-integrin, β7-integrin, or VCAM-1, but not by anti-intercellular adhesion molecule (ICAM)-1. From this observation we can conclude that MAdCAM-1 and VCAM-1, but not ICAM-1, mediate T-lymphocyte recruitment in the inflamed colon. Spatial heterogeneity of MAdCAM-1 and VCAM-1 activation following TNF-α challenge may enhance specific T-lymphocyte recruitment (α4β7 plus α4β1-integrin positive cells, and α4β1-integrin positive cells, respectively) in the inflamed colonic mucosa.

Rolling T cells activate integrins when they receive signals from chemokines on endothelial cell surfaces. Chemokines are secreted polypeptides that bind to specific surface receptors, which transmit signals through G proteins. Recently some chemokines have been shown to trigger intravascular adhesion, whereas others direct the migration of leukocytes into and within the extravascular space. Campbell et al. reported that stromal cell-derived facor-1 (CXCL12), 6-C-kine (CCL21), and macrophage inflammatory protein (MIP)-3β (CCL19) could induce adhesion of most circulating lymphocytes, including most $CD4^+$ T cells, and to induce arrest of rolling cells within 1 s under flow conditions similar to those of blood [10]. They also showed that MIP-3α (CCL20) triggered adhesion of memory, but not naïve $CD4^+$ T cells. Inflammatory cytokines such as RANTES, MIP-1α, and MIP1-β bind a distinct chemokine receptor on Th1 cells, CCR5, but these chemokines may mainly induce these cells to infiltrate into the inflammatory tissues. In Peyer's patches chemotactic stimulation of rolling T cells is known to rely on the pathway involving CCR7, which activates α4β7 integrin and αLβ2 integrin [11].

Recently we found that a CC chemokine, TECK/CCL25, also plays an important role in the adherence of mucosal lymphocytes to the microvessels of the small intestine [12]. TECK/CCL25, which had initially been reported to be produced by thymic cells, is highly expressed at the message level in the small intestine, and has recently been localized to the crypt epithelium. TECK/CCL25 is also known to be expressed by endothelial cells and a subset of cells in the intestinal lamina propria. Interestingly, the only known receptor for TECK/CCL25, CCR9, is expressed by discrete subsets of circulatory memory $CD4^+$ and $CD8^+$ lymphocytes expressing intestinal homing receptor α4β7, suggesting that CCR9 may be a receptor used preferentially by lymphocytes involved in gut immunity. Lamina propria lymphocytes and IELs adhered to the mucosal microvessels of both small intestine and colon, but the blocking of the TECK/CCL25–CCR9 system by desensitization of CCR9 with TECK/CCL25 or anti-TECK/CCL25 antibody treatment significantly inhibited these adhesions only in the small intestine. Desensitization of CCR9 or anti-TECK/CCL25 antibody also attenuated the TNF-α-induced LPL migration in the small intestinal mucosa. These results suggest that the TECK/CCL25–CCR9 system may play a significant role in LPL and IEL adhesion to the microvessels of the small intestine, but not colon, under control as well as inflamed conditions. We speculate that chemokines other than TECK/CCL25 may function in lymphocyte migration in the inflamed colonic mucosa after the administration of TNF-α.

Monocyte Migration in Peyer's Patches and Intestinal Mucosa

Monocyte adhesion to the vascular endothelium and subset diapedesis are important events that occur during chronic inflammation, immune mediated reactions and atherosclerosis. Several monocyte–endothelial adhesion pathways have been described, including adhesion molecules such as L-selectin–peripheral node addressin (PNAd), β2-integrin–ICAM-1, α4-integrin–VCAM-1, and P-selectin glycoprotein-1 (PSGL-1)–P-selectin [13], and subsequent monocyte migration into tissues is regulated by activation of vascular endothelial cells by cytokines such as interleukin (IL)-1, TNF-α, or interferon (IFN)-γ, and production of chemotactic factors including interferon-inducible protein (IP)-10 or monocyte chemoattractant protein (MCP)-1. Although intestinal mucosa monocytes play a pivotal role in specific immunological protection against enteric pathogens, and in chronic inflammatory changes such as Crohn's disease, it has not been clearly understood how circulatory monocytes actually migrate to the intestinal mucosa.

We monitored the migration of fluorescence-labeled monocytes derived from spleen in intestinal microvessels with or without LPS (lipopolysaccharide) treatment to investigate the role of various adhesion molecules (P-selectin, VCAM-1, and ICAM-1) [14]. We found that in control mice without LPS, interactions between infused monocytes and endothelium of small intestinal microvessels were negligible; on the other hand, significant monocyte–endothelial interactions were observed after LPS treatment in postcapillary venules of Peyer's patches, submucosal venules, and microvessels of intestinal villi. Anti-P-selectin monoclonal antibody (mAb) significantly suppressed the LPS-induced increase in monocyte rolling in Peyer's patches and submucosal venules. Anti-VCAM-1 mAb significantly suppressed the LPS-induced monocyte adherent to each site. But anti-ICAM-1 mAb significantly inhibited monocyte adhesion only in postcapillary venules of Peyer's patches, suggesting the more dependency on VCAM-1 in nonlymphoid regions than Peyer's patches at inflammatory conditions.

Role of Adhesion Molecule in Inflammatory Bowel Disease (IBD) and its Animal Models

There is now growing evidence that IBD is associated with altered lymphocyte trafficking in the intestinal mucosa. MAdCAM-1 has been implicated in the selective recruitment of lymphocytes to the gut, and in human ulcerative colitis and Crohn's disease, the expression of MAdCAM-1 is upregulated in factor VIII-positive vessels in inflamed colonic mucosa [15]. Although other adhesion molecules such as VCAM-1 are upregulated in the intestinal mucosa of inflammatory bowel disease, a possible involvement of VCAM-1 in lymphocyte recruitment into the inflamed gut in IBD is still controversial, and the relative functional importance of MAdCAM-1 and VCAM-1 in altered lymphocyte trafficking should be elucidated, especially in the chronic type of colitis.

We investigated this issue using two different models for colitis; one is dextran sulfate sodium (DSS)-induced colitis in mice [16], the other is granulomatous colitis in rats induced by peptidoglycan-polysaccharide (PG-PS) [17]. In DSS colitis BALB/c mice that received 5% DSS solution for 14 days had significant histological damage with shortening of crypts observed in colonic mucosa. Before mucosal inflammation had become significant, expression of MAdCAM-1 was already increased in the microvessels of the lamina propria, preceding the lymphocyte infiltration and VCAM-1 upregulation. Administration of anti-MAdCAM-1 antibody significantly reduced colonic injury as well as the infiltration of β7-integrin-postitive lymphocytes, and this antibody also was effective when given 7 days after the start of DSS treatment. Peptidoglycan-polysaccharide is a structural component of the cell walls of bacteria that has well-described proinflammatory properties. After subserosal intestinal injection, PG-PS induces chronic relapsing inflammation in a susceptible strain (Lewis) of rat. The characteristic inflammatory hallmarks of this model are the development of transmural, granulomatous enterocolitis, which shares several histological features of Crohn's disease. In our study the colonic submucosa was drastically thickened after PG-PS injection with infiltration of CD4-positive lymphocytes and ED-1-positive macrophages. Enhanced MAdCAM-1 expression was demonstrated in venular endothelium of the submucosal venules in inflamed colon. Administration of anti-MAdCAM-1 antibody (OST-2) significantly attenuated the PG-PS-induced colonic damage and cell infiltration, while nonblocking antibody (OST-20) did not. The attenuating effect of anti-MAdCAM-1 suggests the importance of the MAdCAM-1 dependent process in the formation of chronic granulomatous colitis.

MAdCAM-1 is induced on a murine endothelial cell line by TNF-α and IL-1-mediated nuclear factor κB protein in vitro. Because MAdCAM-1 is an adhesion molecule more unique to the gut mucosa than other adhesion molecules such as VCAM-1 or ICAM-1, attenuation of MAdCAM-1 may be more specific therapy for colitis. However, as shown in our studies, MAdCAM-1 is also essential for lymphocyte trafficking to gut-associated lymphoid tissues as well as to the intestinal mucosa under physiological conditions. It was also suggested that MAdCAM-1 functions as a vascular addressin in not only mucosal, but also nonmucosal lymphoid tissues during ontogeny [18]. Although it has not been clinically applied yet, it should be taken into account not to oversuppress MAdCAM-1 function if anti-MAdCAM-1 blocking is used for treatment of IBD. Recently the European study group described results of a double-blinded, randomized trial comparing natalizumab, a recombinant humanized monoclonal antibody, against α4 integrin, with placebo in the treatment of moderate-to-severe Crohn's disease [19]. In this trial, natalizumab induced remissions and responses roughly twice as often as placebo. The authors concluded that in the short-term study natalizumab showed efficacy for reducing signs and symptoms at least similar to that of infliximab. Treatment with selective adhesion-molecule inhibitors may become a potential strategy for management of intestinal inflammation in future.

References

1. Bargatze RF, Jutila MA, Butcher EC (1995) Distinct roles of L-selectin and integrins α4β7 and LFA-1 in lymphocyte homing to Peyer's patch-HEV in situ: The multistep model confirmed and refined. Immunity 3:99–108
2. Fujimori H, Miura S, Koseki S, et al (2002) Intravital observation of adhesion of lamina propria lymphocytes to microvessels of small intestine in mice. Gastroenterology 122:734–744
3. Miura S, Tsuzuki T, Fukumura D, et al (1995) Intravital demonstration of sequential migration process of lymphocyte subpopulations in rat Peyer's patches. Gastroenterology 109:1113–1123
4. Miura S, Serizawa H, Tsuzuki Y, et al (1997) Vasoactive intestinal peptide modulates T lymphocyte migration in Peyer's patches of rat small intestine. Am J Physiol Gastrointest Liver Physiol 272:G92–G99
5. Hokari R, Miura S, Fujimori H, et al (1998) Nitric oxide modulates T-lymphocytes migration in Peyer's patches and villous submucosa of rat small intestine. Gastroenterology 115:618–627

6. Tsuzuki Y, Miura S, Kurose I, et al (1997) Enhanced lymphocyte interaction in post-capillary venules of Peyer's patches during fat absorption in rats. Gastroenterology 112:813–825
7. Wagner N, Löhler J, Kunkel EJ, et al (1996) Critical role for β7 integrin in formation of the gut-associated lymphoid tissue. Nature 382:366–370
8. Koseki S, Miura S, Fujimori H, et al (2001) In situ demonstration of intraepithelial lymphocyte adhesion to villus microvessels of the small intestine. Int Immunol 13:1165–1174
9. Watanabe C, Miura S, Hokari R, et al (2002) Spatial heterogeneity of TNF-alpha-induced T cell migration to colonic mucosa is mediated by MAdCAM-1 and VCAM-1. Am J Physiol Gastrointest Liver Physiol 283:G1379–G1387
10. Campbell JJ, Hedrick J, Zlotnik A, et al (1998) Chemokines and the arrest of lymphocytes rolling under flow conditions. Science 279:381–384
11. Warnock RA, Campbell JJ, Dorf ME, et al (2000) The role of chemokines in the microenvironmental control of T versus B cell arrest in Peyer's patch high endothelial venules. J Exp Med 191:77–88
12. Hosoe N, Miura S, Watanabe C, et al (2004) Demonstration of functional role of TECK/CCL25 in T lymphocyte-endothelium interaction in inflamed and uninflamed intestinal mucosa. Am J Physiol Gastrointest Liver Physiol 286:G458–G466
13. McEvoy LM, Julita MA, Tsao PS, et al (1997) Anti-CD43 inhibits monocyte-endothelial adhesion in inflammation and atherogenesis. Blood 90:3587–3594
14. Ishii N, Tsuzuki Y, Matsuzaki K, et al (2004) Endotoxin stimulates monocyte-endothelial cell interactions in mouse intestinal Peyer's patches and villus mucosa. Clin Exp Immunol 135:226–232
15. Briskin MJ, Winsor-Hines D, Shyjan A, et al (1997) Human mucosal addressin cell adhesion molecule-1 is preferentially expressed in intestinal tract and associated lymphoid tissue. Am J Pathol 151:97–110
16. Kato S, Hokari R, Matsuzaki K, et al (2000) Amelioration of murine experimental colitis by inhibition of mucosal addressin cell adhesion molecule-1. J Pharmacol Exp Ther 295:183–189
17. Hokari R, Kato S, Matsuzaki K, et al (2001) Involvement of mucosal addressin cell adhesion molecule-1 (MAdCAM-1) in the pathogenesis of granulomatous colitis in rats. Clin Exp Immunol 126:259–265
18. Iizuka T, Tanaka T, Suematsu M, et al (2000) Stage-specific expression of mucosal addressin cell adhesion molecule-1 during embryogenesis in rats. J Immunol 164: 2463–2471
19. Ghosh S, Goldin E, Gordon FH, et al (2003) Natalizumab Pan-European Study Group. Natalizumab for active Crohn's disease. N Engl J Med 348:24–32

Role for P-Selectin in Platelet Thrombus Formation Under Flow

Makoto Handa

P-selectin (PS) is a membrane glycoprotein existing in α granules of the platelets and Weibel–Palade bodies of vascular endothelial cells. Following cellular activation, it becomes surface expressed by exocytosis. PS mediates leukocyte rolling at inflamed or injured sites of vasculature via interactions with its counter receptor PSGL-1. Likewise, previous studies suggested its role as a target molecule of platelet rolling on stimulated endothelium by binding with platelet GP Ib/V/IX complex.

Platelet adhesion to and subsequent platelet–platelet interactions with perturbed vascular wall is a prerequisite for hemostasis and thrombosis. Since adhered platelets may provide circulating platelets with a PS-rich surface, it is speculated that the glycoprotein may have a role in platelet thrombus formation under flow. To test the above notion, we have used an in vitro flow system consisting of a laminar flow chamber equipped with an epifluorescence microscope. PPACK-treated whole blood was recirculated over a platelet surface that was made by incubating washed platelets with a collagen-coated glass slip. Adhesive interactions of fluorescence-labeled platelets with the surface were captured using a CCD camera and were analyzed by an image analyzer.

Platelet adhesion to the platelet surface (surface coverage) at lower flow rates (≦1200/s) was partially but readily inhibited (inhibition: 36.7% ± 3.6% at 600/s, 80 s, $n = 3$) by a function-blocking anti-rat PS mouse monoclonal antibody S789G. In contrast, the inhibitory effect of anti-von Willebrand factor antibody NMC-4 was dependent upon flow rate (50.3% ± 1.7% at 600/s, 95% ± 3.5% at 2400/s). At a flow rate of 600/s, the two antibodies in combination blocked platelet adhesion by 84% ± 2.2% that closely approximated a level obtained by anti-GPIb antibody GUR83-35 (90.5% ± 3.4%). The effect of

Blood Center, Keio University School of Medicine, 35 Shinanomachi, Shinjuku-ku, Tokyo 160-8582, Japan

S789G was found to be due mainly to the blockade of platelet rolling on the surface. Experiments using a PS-coated surface (PS was purified from a soluble platelet membrane fraction produced by sonication) demonstrated that PS-mediated platelet rolling under lower flow conditions depended on GP Ib/V/IX.

These results suggest that PS on activated platelet monolayer may mediate initial contacts as a coligand with vWF for GP Ib/IX on flowing platelets under venous flow conditions.

Glomerular Microcirculation: Distinct Intracellular Mechanisms for Afferent and Efferent Arteriolar Tone

Koichi Hayashi, Koichiro Homma, Shu Wakino, Tsuneo Takenaka, Hiroo Kumagai, and Takao Saruta

Summary. Angiotensin (ANG) II contributes importantly to the regulation of renal pre- and postglomerular arteriolar tone. The present study examined the subcellular signaling mechanisms for ANG II-induced afferent (AFF) and efferent arteriolar (EFF) constriction, using the isolated perfused hydronephrotic rat kidney. Angiotensin II-induced AFF and EFF constriction was abolished by an ANG II receptor antagonist (losartan). The pretreatment with *N*-ethylmaleimide (G_i protein inhibitor) completely prevented the ANG II-induced constriction of EFF, but not AFF. Furthermore, signal interruption at the phospholipase C level by 2-nitro-4-carboxyphenyl-*N,N*-diphenyl-carbamate blocked the constriction of both arterioles. Next, ANG II-induced AFF constriction was completely inhibited by the blockade of inositol-1,4,5-trisphosphate (IP_3) signaling by thapsigargin and L-type voltage-dependent calcium channel blockers, but relatively refractory to protein kinase C (PKC) inhibition (by chelerythrine). In contrast, EFF constriction was resistant to pranidipine, but partially responsive to thapsigargin and chelerythrine. Finally, direct PKC activation by phorbol myristate acetate caused prominent EFF constriction, which was inhibited by manganese/free calcium medium, but not by pranidipine. Thus, PKC plays an obligatory role in ANG II-induced EFF constriction that requires extracellular calcium entry through non-selective cation channels. By contrast, ANG II-induced AFF constriction is mainly mediated by IP_3 and voltage-dependent calcium channel pathways. Collectively, intracellular signaling mechanisms differ in AFF and EFF, which may determine the glomerular function.

Key words. Afferent arteriole, Efferent arteriole, Calcium channel, Inositol trisphosphate, Protein kinase C

Department of Internal Medicine, Keio University School of Medicine, 35 Shinanomachi, Shinjuku-ku, Tokyo 160-8582, Japan

Introduction

The kidney serves to function as an important organ to maintain body fluid homeostasis as well as electrolyte balance. Thus, both preglomerular afferent and postglomerular efferent arterioles play a pivotal role to govern glomerular capillary pressure, and subsequently the glomerular filtration rate (GFR). However, the regulation of afferent and efferent arteriolar tone should teleologically differ because parallel changes in the arteriolar resistance of these vessels would not alter GFR. Indeed, histological characteristics differ between afferent and efferent arterioles. Thus, vascular smooth muscles of afferent arterioles manifest spindle-shaped morphology, whereas those of efferent arterioles possess star-like morphology [1]. Similarly, myosin heavy chain isoform SM1 is reported to be present in both afferent and efferent arterioles, whereas SM2 is present only in the afferent arteriole [2]. Furthermore, although it is established that angiotensin (ANG) II is abundant in the kidney and represents an important intrarenal paracrine regulating the vascular tone of pre- (afferent) and post-glomerular (efferent) arterioles [3], the functional heterogeneity in the renal microvascular action of ANG II remains fully undetermined. Furthermore, whether intracellular signaling pathways differ in afferent and efferent arterioles is not elucidated.

Direct Visualization of Renal Microcirculation

To clarify the different responsiveness of afferent and efferent arterioles to ANG II, we examined the arteriolar behavior, using the isolated perfused hydronephrotic rat kidney model that directly visualized the renal microvessels in vitro [4–6]. The rat kidney was rendered hydronephrosis by ligation of the right ureter. After 6–8 weeks, renal tubular atrophy had progressed that allowed direct visualization of the renal microcirculation.

Figure 1 illustrates a schematic diagram of the apparatus used in the present study. The hydronephrotic kidney was placed on the stage of an inverted microscope modified to accommodate a heated chamber equipped with a thin glass viewing port on the bottom surface. Kidneys were perfused with medium saturated with a gas mixture of O_2/CO_2 within a pressurized reservoir. The perfusion pressure was controlled by adjusting the back-pressure-type regulator. Images were obtained from an inverted microscope, and were captured by a computer equipped with a video acquisition board.

Using this experimental model, we found that the responsiveness of afferent and efferent arterioles varied, depending on the underlying vasoconstrictors used. Thus, as shown in Fig. 2, efferent arteriolar responses to

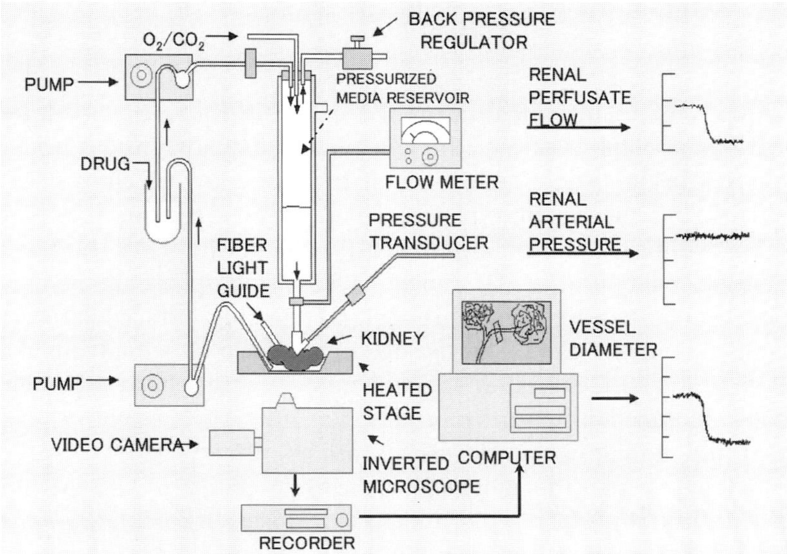

FIG. 1. Apparatus used to study microvessels in the isolated perfused hydronephrotic kidney. The hydronephrotic kidney was placed on the stage of an inverted microscope modified to accommodate a heated chamber equipped with a thin glass viewing port on the bottom surface. Kidneys were perfused with medium saturated with a gas mixture of O_2/CO_2 within a pressurized reservoir. The perfusion pressure was controlled by adjusting the back-pressure-type regulator. Images were obtained from an inverted microscope, and were captured by a computer equipped with a video acquisition board

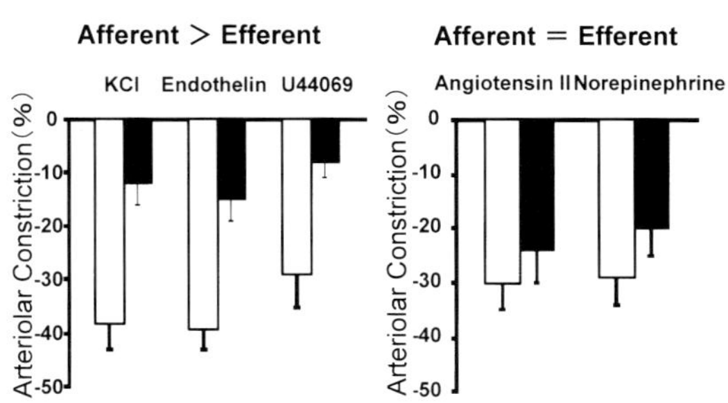

FIG. 2. Distinct responsiveness of afferent and efferent arterioles to vasoconstrictor stimuli. *Open bars*, afferent arterioles; *filled bars*, efferent arterioles

these agents were blunted, as compared with those of afferent arterioles. In contrast, both ANG II and norepinephrine caused similar magnitude of vasoconstriction of afferent and efferent arterioles. Of note, it is demonstrated that similar magnitude of constriction of afferent and efferent arterioles results in an elevation in glomerular capillary pressure, leading to an increase in GFR [7].

Intracellular Signaling of ANG II

To delineate the cellular mechanism of the ANG II-induced constriction of afferent and efferent arterioles, the following signaling pathways have been examined. Thus, in general ANG II binds to its receptor, which couples with G proteins, and subsequently phospholipase C. This enzyme breaks down phosphatidylinositol 4,5-bisphosphate (PIP_2) into inositol 1,4,5-trisphosphate (IP_3) and diacylglycerol (DAG). Inositol 1,4,5-trisphosphate then stimulates IP_3 receptors at the sarcoplasmic reticulum, releasing intracellular Ca from the sarcoplasmic reticulum. The elevated intracellular Ca would stimulate Cl channels and cause membrane depolarization, opening voltage-dependent Ca channels. On the other hand, the formation of DAG stimulates protein kinase C (PKC), and modifies smooth muscle contraction.

We therefore tested the above-mentioned hypothesis in the renal microcirculation. Initially, AT1 receptor blockade by losartan completely inhibited the ANG II-induced constriction of both afferent and efferent arterioles in a dose-dependent manner [8]. These observations indicate that AT1 receptors constitute an important determinant that mediates the ANG II-induced constriction of renal microvessels. Next, a G_i protein inhibitor, *N*-ethylmaleimide, completely abolished the ANG II-induced efferent arteriolar constriction. This agent however failed to alter the afferent arteriolar constriction. Furthermore, a phospholipase C inhibitor, 2-nitro-4-carboxyphenyl-*N,N*-diphenyl carbamate (NCDC), completely blocked the ANG II-induced constriction of both afferent and efferent arterioles [9]. Collectively, ANG II receptor stimulation would involve G protein and PLC in both afferent and efferent arterioles. However, at the G protein level, G_i protein participates in the ANG II-induced efferent arteriolar tone, whereas other G protein, such as G_q, may be responsible for afferent arteriolar constriction induced by ANG II.

Next, subcellular mechanisms for ANG II-induced constriction were evaluated, including IP_3 and DAG. In the presence of thapsigargin, which finally inhibits the IP_3-mediated Ca release from the sarcoplasmic reticulum, the afferent arteriolar response to ANG II was completely abolished, whereas the efferent arteriolar response was diminished but not totally abolished [9] (Fig. 3). Furthermore, a chloride channel blocker, IAA-94, reversed the ANG II-induced constriction of the afferent arteriole in a dose-dependent manner [10]. Similarly, a voltage-dependent Ca channel blocker, nifedipine, prominently reversed the ANG II-induced constriction of the afferent, but not efferent, arteriole [11].

In contrast, the pretreatment with staurosporine had modest effect on the ANG II-induced constriction of the afferent arteriole, whereas the efferent

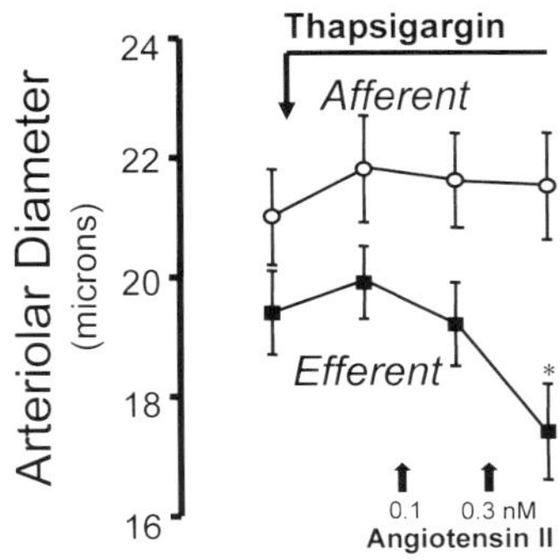

FIG. 3. Role of intracellular Ca release in angiotensin II-induced constriction of renal microvessels. In the presence of thapsigargin, angiotensin II failed to alter afferent arteriolar diameter, but elicited constriction of efferent arterioles. $*P < 0.05$ vs baseline

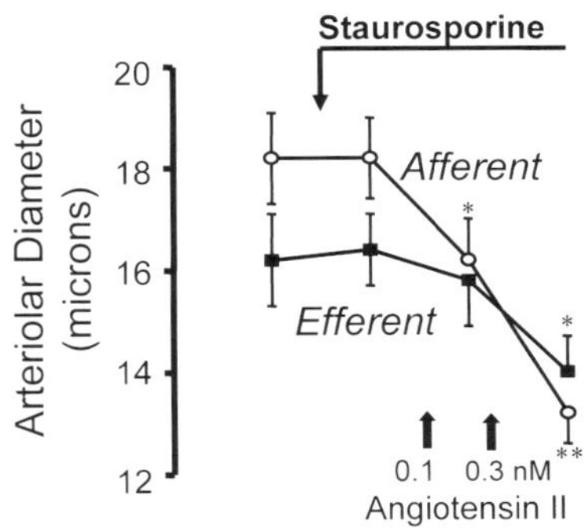

FIG. 4. Protein kinase C contributes more greatly to efferent than afferent arteriolar tone during angiotensin II-induced stimulation. $*P < 0.05$ vs baseline; $**P < 0.01$ vs baseline

arteriolar response was partially prevented by staurosporine [10] (Fig. 4). In concert, the above-mentioned observations on the afferent arteriole indicate that the IP_3 formed by phospholipase C stimulates Ca release from sarcoplasmic reticulum (Fig. 5). The increased Ca would enhance Cl channel activity and the subsequent membrane depolarization. Such elevated membrane potential gates voltage-dependent Ca channels and finally induces Ca influx from this channel.

Next, we focus on the role of the intracellular Ca release mechanism in mediating the ANG II-induced efferent arteriolar constriction. As shown in Fig. 6, thapsigargin potently inhibited the ANG II-induced constriction [10]. Of note, the vasoconstrictor component was still retained in the presence of thapsigargin. Similarly, ANG II-induced constriction was more sensitive to the inhibition of PKC by chelerythrine in the efferent than in the afferent arteriole [12], thus suggesting that the PKC-mediated vasoconstrictor tone of the efferent arteriole was more greatly activated in the efferent arteriole.

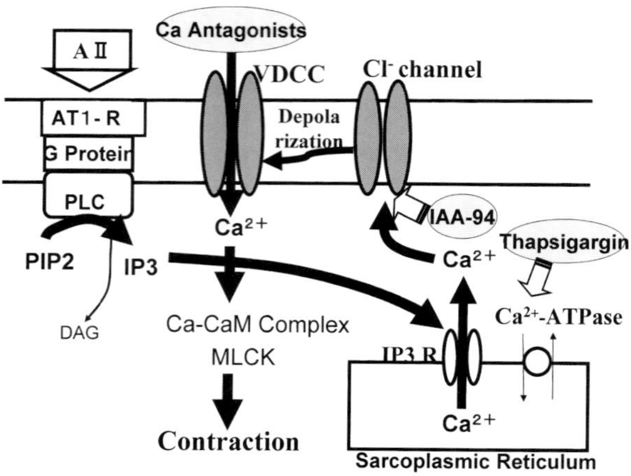

FIG. 5. Mechanism for angiotensin II-induced constriction of the afferent arteriole. *AT1*, angiotensin type 1 receptors; *VDCC*, voltage-dependent Ca channels; *PLC*, phospholipase C; *DG*, diacylglycerol; *PKC*, protein kinase C; *MLCK*, myosin light chain kinase; *CaM*, calmodulin

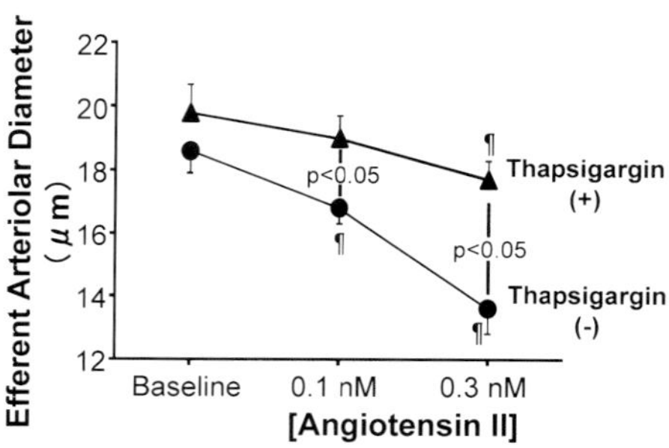

FIG. 6. Role of intracellular Ca release in angiotensin II-induced efferent arteriolar constriction. ¶$P < 0.05$ vs baseline

Role of Intracellular Ca

When extracellular Ca is removed, afferent arterioles failed to constrict in response to ANG II. In the efferent arteriole, however, removing extracellular Ca or addition of manganese diminished, but not completely abolished the efferent arteriolar constriction [10].

To clarify whether extracellular Ca is involved in the PKC-mediated efferent arteriolar constriction, we examined the effect of removing extracellular Ca on phorbol myristate acetate (PMA)-induced constriction of the efferent arteriole. Thus, this manipulation completely inhibited the PMA-induced vasoconstriction of this vessel [12].

In concert, in the efferent arteriole, after binding with its receptors, ANG II-stimulated IP_3 would release Ca from sarcoplasmic reticulum, and could involve Ca influx from store-operated Ca channels (Fig. 7). In parallel with this mechanism, ANG II enhances PKC, which opens nonselective cation channels, or TRP channels, and subsequently elevates intracellular Ca concentration.

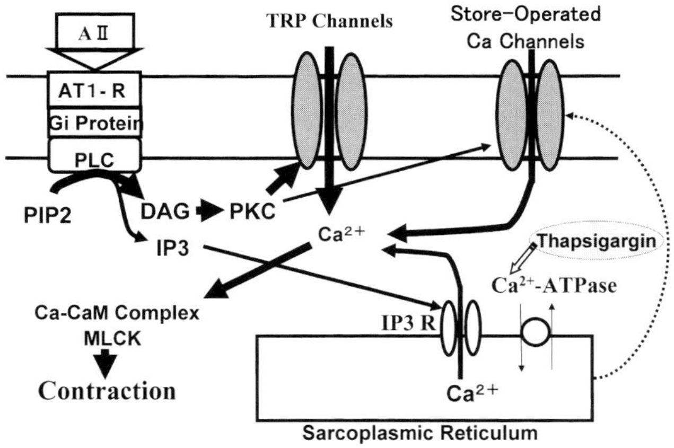

FIG. 7. Mechanism for angiotensin II-induced constriction of efferent arterioles. *AT1*, angiotensin type 1 receptors; *TRP*, transient receptor protein potential; *PLC*, phospholipase C; *DAG*, diacylglycerol; *PKC*, protein kinase C; *MLCK*, myosin light chain kinase; *CaM*, calmodulin

TABLE 1. Summary showing different mechanisms for angiotensin II-induced constriction of renal microvessels

	Afferent		Efferent
1. AT1 receptors		=	
2. G protein	? (G_q)		G_i
3. Phospholipase C		=	
4. IP3/DAG-PKC	IP3 ≫ PKC		IP3 < PKC
5. ECF Ca-dependency	++		+
6. Ca influx pathway	VDCC		NSCC

AT1, angiotensin type 1 receptors; PKC, protein kinase C; DAG, diacylglycerol; VDCC, voltage-dependent Ca channels; NSCC, nonselective cation channels

Concluding Remarks

Although abundance of the intrarenal ANG II content clearly suggests an important role of this substance in the regulation of renal function as a determinant of glomerular hemodynamics, substantial evidence has accrued that ANG II elicits distinct activity on afferent and efferent arterioles. The IP_3-mediated signaling pathway constitutes an important determinant of the afferent arteriolar response to ANG II (Table 1). In contrast, both IP_3 and DAG-PKC pathways contribute to the development of ANG II-induced constriction of the efferent arteriole. In conclusion, the roles of intracellular signaling mechanisms, including G_i protein, voltage-dependent calcium channels, IP_3, and PKC, during ANG II stimulation differ in afferent and efferent arterioles, which may constitute segmental heterogeneity in the renal microvasculature, and therefore would serve to control the glomerular capillary pressure and subsequent renal injury.

References

1. Gattone VH 2nd, Luft FC, Evan AP (1984) Renal afferent and efferent arterioles of the rabbit. Am J Physiol 247:F219–F128
2. Kimura K, Nagai R, Sakai T, et al (1995) Diversity and variability of smooth muscle phenotypes of renal arterioles as revealed by myosin isoform expression. Kidney Int 48:372–382
3. Navar LG, Inscho EW, Majid SA, et al (1996) Paracrine regulation of the renal microcirculation. Physiol Rev 76:425–536
4. Loutzenhiser R, Hayashi K, Epstein M (1988) Atrial natriuretic peptide reverses afferent arteriolar vasoconstriction and potentiates efferent arteriolar vasoconstriction in the isolated perfused rat kidney. J Pharmacol Exp Ther 246:522–528
5. Hayashi K, Epstein M, Loutzenhiser R (1989) Pressure-induced vasoconstriction of renal microvessels in normotensive and hypertensive rats: Studies in the isolated perfused hydronephrotic kidney. Circ Res 65:1475–1484
6. Hayashi K, Epstein M, Loutzenhiser R (1990) Determinants of renal actions of atrial natriuretic peptide: lack of effect of atrial natriuretic peptide on pressure-induced vasoconstriction. Circ Res 67:1–10
7. Carmines PK, Perry MD, Hazelrig JB, et al (1987) Effects of preglomerular and postglomerular vascular resistance alterations on filtration fraction. Kidney Int Suppl 20:S229–S232
8. Loutzenhiser R, Epstein M, Hayashi K, et al (1991) Characterization of the renal microvascular effects of angiotensin II antagonist, DuP753: studies in isolated perfused hydronephrotic kidneys. Am J Hypertens 4:309s–314s
9. Takenaka T, Suzuki H, Kanno Y, et al (1997) Cellular mechanisms mediating rat renal microvascular constriction by angiotensin II. J Clin Invest 100:2107–2114
10. Takenaka T, Kanno Y, Kitamura Y, et al (1996) Role of chloride channels in afferent arteriolar constriction. Kidney Int 50:864–872
11. Hayashi K, Nagahama T, Oka K, et al (1996) Disparate effects of calcium antagonists on renal microcirculation. Hypertens Res 19:31–36
12. Nagahama T, Hayashi K, Ozawa Y, et al (2000) Role of protein kinase C in angiotensin II-induced constriction of renal microvasculature. Kidney Int 57:215–223

Role of Lectin-Like Oxidized Low-Density Lipoprotein Receptor-1 (LOX-1) in Adenosine Diphosphate-Activated Platelet Adhesion to Human Brain Microvascular Endothelial Cells Under Flow In Vitro

Norio Tanahashi[1], Yoshiaki Itoh[1], Hideki Satoh[1], Takato Abe[1], Hidetaka Takeda[1], Minoru Tomita[1], and Tatsuya Sawamura[2]

Key words. Platelet adhesion, Human brain microvascular endothelial cell, Lectin-like oxidized low-density lipoprotein receptor-1, Video-enhanced contrast microscopy, Adenosine diphosphate

Introduction

We previously observed that activated platelets adhered to human brain microvascular endothelial cells (HBEC) under flow in vitro using video enhanced contrast (VEC) microscopy [1], and we showed that a platelet glycoprotein (GP)IIb/IIIa antagonist ameliorated platelet adhesion to HBEC [2]. The molecular mechanism of platelet adhesion to endothelial cells is still not fully clarified. Recently, lectin-like oxidized low-density lipoprotein receptor-1 (LOX-1) was identified in endothelial cells [3] and platelets [4]. It was also found that negative phospholipids exposed at the surface of platelets on activation are epitopes of LOX-1 [5], and LOX-1 could work as an adhesion molecule for platelets. The purpose of the present study was to examine the role of LOX-1 in platelet adhesion to HBEC.

[1]Department of Neurology, Keio University School of Medicine, 35 Shinanomachi, Shinjuku-ku, Tokyo 160-8582, Japan
[2]National Cardiovascular Center Research Institute, Suita, Osaka 565-8566, Japan

Methods

Human brain microvascular endothelial cells in the third passage were purchased (Applied Cell Biology Research Institute, USA). They were cultured with medium (HuMedia-EG, Kurabo, Japan) and at the end of the fourth passage, the cells were disseminated at a concentration of 5000 cells/cm^2 on glass slides coated with collagen. After 5–7 days, when the cells were confluent, they were used for the experiment. Platelet-rich plasma (PRP) and platelet-poor plasma (PPP) were prepared from venous blood from normal adult volunteers. Platelet-rich plasma was diluted with PPP to give a final concentration of 300 000/μl. To evaluate platelet adhesion at high magnification, we employed a VEC microscopy system, which consisted of an inverted Nomarski microscope (Axiovert 135, Carl Zeiss, Germany), objective lenses for differential interference contrast, insertion lenses, and a halogen lamp. For image analysis, a charge-coupled device (CCD) camera (C3077, Hamamatsu Photonics, Japan), an image analyzer (Argus-10, Hamamatsu Photonics), and an S-VHS video recorder were used. This VEC microscopy system enabled us to observe cells on a slide glass at a maximum magnification of ×12 000. On the microscope stage, a slide glass with endothelial cells in an observation chamber was mounted. Perfusate was pumped from one end to fill the chamber. The drainage was collected from the other end, so that perfusate flowed over the endothelial cells at a constant rate. To quantify the degree of platelet adhesion to HBEC, the number of adhering platelets in a field 30 μm in length and breadth was counted. One hundred consecutive fields (10 × 10) were counted and the average number was calculated. In the case of platelet clusters, only the superficial platelets were counted. A large mass of more than 100 platelets was counted as 100 platelets. In the control group ($n = 7$), HBECs were cultured on a coverglass and put in the observation chamber for VEC microscopy. Then, PRP was superfused with an infusion pump at a low shear rate (10/s) for 30 min and washed out. Interaction between platelets and endothelial cells was observed by VEC microscopy and the number of platelets that adhered to HBEC was calculated. In the adenosine diphosphate (ADP) group ($n = 7$), PRP with ADP (2 μM) was superfused for 30 min and platelet adhesion to HBEC was observed. In the anti-LOX-1 group ($n = 5$), PRP with ADP (2 μM) plus JTX92 (10 μg/ml; antihuman LOX-1 antibody) was superfused for 30 min and platelet adhesion to HBEC was observed.

Results

In the control group, platelet adhesion to HBEC was rarely seen. However, ADP-stimulated platelets adhered to HBEC (Fig. 1). As shown in Fig. 2, antihuman LOX-1 antibody inhibited adhesion of ADP-activated platelets to

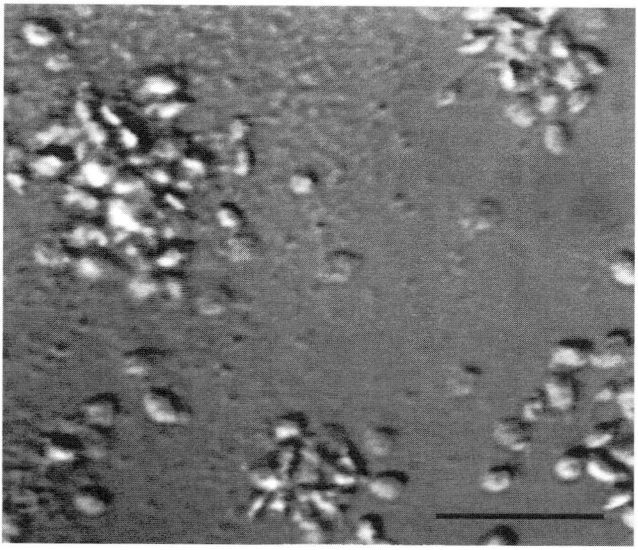

Fig. 1. The adhesion of platelets on HBEC was clearly seen in the adenosine diphosphate (ADP) group. Microaggregates of platelets were also seen. Scale bar = 10 μm

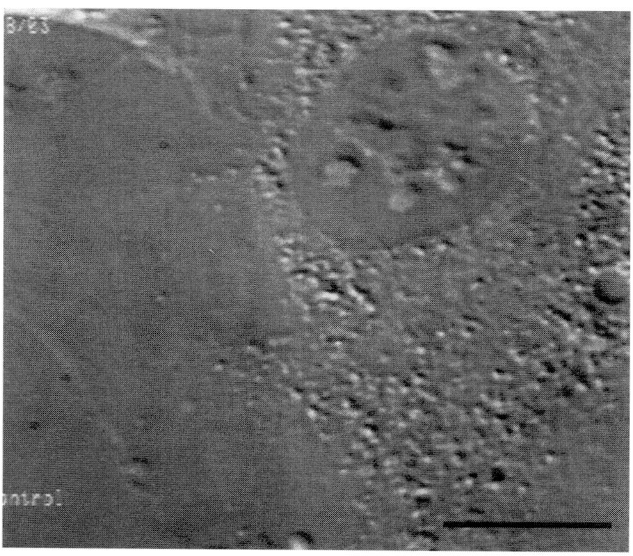

Fig. 2. Platelet adhesion was rarely seen in the anti-LOX-1 (lectin-like oxidized low-density lipoprotein receptor-1) antibody group. Scale bar = 10 μm

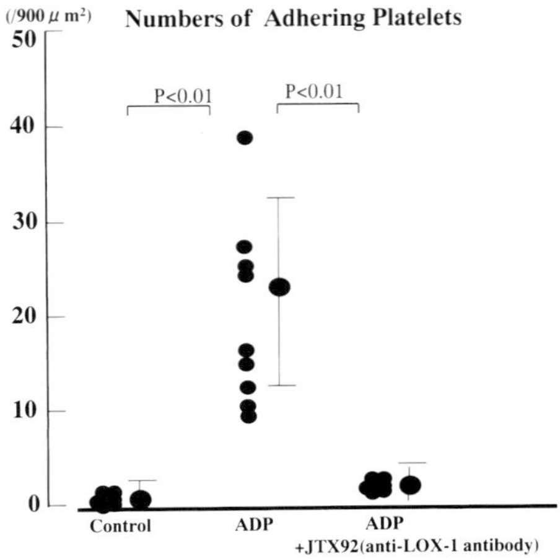

FIG. 3. Number of platelets adhering to endothelial cells after washout in each group. The average numbers of adhering platelets in the anti-LOX-1 antibody group was significantly ($P < 0.01$ by Student's *t*-test) smaller than in the ADP group

HBEC. Figure 3 shows a comparison of the number of adhering platelets among groups. The average number of platelets adhering and aggregating to endothelial cells was $0.3 \pm 0.6/900\,\mu m^2$ in the control group, $22.3 \pm 10.4/900\,\mu m^2$ in the ADP group ($P < 0.01$ vs control group), and $1.8 \pm 2.5/900\,\mu m^2$ in the antihuman LOX-1 antibody group ($P < 0.01$ vs ADP group).

Discussion

The present study showed that LOX-1 is closely related to the molecular mechanism of activated platelet adhesion to HBEC under flow in vitro, and might work as an adhesion molecule for platelets. We previously reported that GPIIb/IIIa receptor and P-selectin play important roles in platelet adhesion to HBEC [2,6]. We consider LOX-1 might be a third mechanism for activated platelet adhesion to HBEC under flow. Upon activation, platelets could adhere to HBEC through GPIIb/IIIa receptor, P-selectin or LOX-1. Kakutani et al [5] verified that negatively charged phospholipids exposed on the surface of activated platelets are epitopes for LOX-1, and Shimaoka et al [7] found that LOX-1 supports cell adhesion to fibronectin in an integrin-independent manner. We need further studies to clarify in detail the role of LOX-1 in the molecular mechanisms of activated platelet adhesion to HBEC. Antihuman LOX-1 antibody significantly reduced activated platelet adhesion to HBEC at a low-flow state in vitro, indicating that LOX-1 plays an important role in activated platelet adhesion to HBEC.

References

1. Tanahashi N, Fukuuchi Y, Tomita M, et al (1999) Platelet adhesion to human brain microvascular endothelial cells in vitro. Observation with video-enhanced contrast microscopy. Neurosci Lett 274:199–202
2. Tanahashi N, Fukuuchi Y, Tomita M, et al (2001) Adhesion of adnosine diphosphate-activated platelets to human brain microvascular endothelial cells under flow in vitro is mediated via GPIIb/IIIa. Neurosci Lett 301:33–36
3. Sawamura T, Kume N, Aoyama T, et al (1997) An endothelial receptor for oxidized low-density lipoprotein. Nature 386:73–77
4. Chen M, Kakutani M, Naruko T, et al (2001) Activation-dependent surface expression of LOX-1 in human platelets. Biochem Biophys Res Commun 282:153–158
5. Kakutani M, Masaki T, Sawamura T (2000) A platelet-endothelium interaction mediated by lectin-like oxidized low-density lipoprotein recptor-1. Proc Natl Acad Sci USA 97:360–364
6. Abe T, Tanahashi N, Satoh H, et al (2001) Role of P-selectin in the adhesion of adenosine diphosphate-activated platelets to human brain microvascular endothelial cells under in vitro flow condition. In: Tsuchiya M, Asano M, Kajiya F (eds) Microcirculation annual 2001. Nihon-Igakukan, pp 83–84
7. Shimaoka T, Kume N, Minami M, et al (2001) Lectin-like oxidized low density lipoprotein receptor-1 (LOX-1) supports cell adhesion to fibronectin. FEBS Lett 504:65–68

Role of Oxidative Stress in Interaction Between Endothelial Cells and Platelets in Diabetes

YASUNORI HIROSE[1], EIICHI SEKIZUKA[3], HIROMICHI NAKADATE[1], TADASHI OZAWA[1], HARUYUKI MINAMITANI[1], CHIKARA OSHIO[4], and HIROMASA ISHII[2]

Key words. Diabetes mellitus, Endothelial cell, Platelet, Advanced glycation end product, Oxidative stress

Introduction

Diabetes mellitus is characteristic of chronic hyperglycemia, and causes various angiopathies. The population of diabetic patients continues to increase, so it is important to clarify the mechanisms of angiopathies and to apply the results to treatment strategies.

The disorders of the circulating system in diabetes mellitus are categorized into macroangiopathy, coronary heart disease, cerebral infarction or arteriosclerosis, and microangiopathy, nephropathy, retinopathy, and neuropathy. Among major factors reported are decreased erythrocyte deformability [1], increased leukocyte chemotaxis, abnormal platelet adherence and aggregation [2], and increased plasma viscosity. However, the mechanisms of these dysfunctions are not clarified as yet, especially in microangiopathy.

In previous studies, we investigated thrombus formation in the mesenteric arterioles of diabetic rats using a photochemical reaction. We confirmed that there was a shortening of thrombus formation times in diabetic rats, but we are unable to clarify the main contributors to thrombus formation in diabetes

[1] Graduate School of Science and Technology, Keio University, 14-1 Hiyoshi 3-chome, Kohoku-ku, Yokohama, Kanagawa 223-0061, Japan
[2] Department of Internal Medicine, Keio University School of Medicine, 35 Shinanomachi, Shinjuku-ku, Tokyo 160-8582, Japan
[3] Department of Clinical Research, National Saitama Hospital, 2-1 Suwa, Wako, Saitama 351-0102, Japan
[4] Oshio Clinic, 2-6-2 Kanda Awajicho, Chiyoda-ku, Tokyo 101-0063, Japan

mellitus, platelets or endothelial cells (ECs). Furthermore, it has been suggested in recent studies that radical oxygen levels rise in diabetic conditions and injure the vessels, leading to some complications. Therefore, we herein investigate the interaction of platelets and ECs under diabetic conditions and the contribution of radical oxygen species (ROS).

Material and Methods

Human umbilical vein ECs were cultured in high D-glucose or N-epsilon-(carboxymethyl)lysine-modified advanced glycation end-products (CML-AGE) in culture solution to assume diabetic conditions. Endothelial cells were kept in culture until they became confluent. We used human healthy platelets and diabetic platelets (HbA1c > 6.5). Blood was drawn from healthy or diabetic donors and mixed with 3.8% sodium citrate at a ratio of 9:1 and centrifuged at $100 \times g$ for 10 min to make platelet-rich plasma. Platelet-rich plasma was diluted with self-plasma to adjust platelet concentration to 300 000/μl.

After culturing ECs in various conditions, Photofrin as the photosensitizer was added to the culture solution and incubated for 30 min. Platelet-rich plasma was brought into contact with incubated ECs, and we irradiated the excitation light and damaged ECs. After light irradiation, we measured the platelet adhesion area (PAA) on ECs by NIH imaging.

To study the effect of ROS for platelet adhesion to ECs, we incubated superoxide dismutase (SOD), catalase, or dimethyl sulfoxide with high D-glucose or high CML-AGE. In this case, we did not use the photochemical reaction in order to eliminate the influence of ROS generated by photochemical reaction. Platelet-rich plasma was brought into contact with ECs for 30 min without photochemical reaction. The method used to calculate PAA was the same as described above.

Results

Platelet adhesion area changed according to the diabetic conditions. When ECs were cultured with high glucose or high CML-AGE, PAA increased significantly compared to normal culture conditions. Platelet adhesion area in diabetic platelets was not different to that from normal platelets under any condition.

Platelet adhesion area was reduced by ROS scavengers in high glucose or high AGE. Radical oxygen species scavengers reduced the increase of PAA by high glucose or high AGE, but did not restore it to the normal level. Suppression of platelet adhesion by H_2O_2 and $^{\cdot}OH$ scavenger was stronger in high CML-AGE than in high D-glucose.

Discussion

Platelet adhesion area increases with diabetic ECs but not with diabetic platelets. So it can be said that ECs under diabetic conditions contributed much more to platelet adhesion than diabetic platelets. This suggests that easy thrombus formation in diabetes mellitus is mainly caused by the dysfunction of ECs.

Platelet adhesion area reduction ratios by various ROS scavengers were calculated based on increased PAA in high glucose or high CML-AGE. The result indicates that the damage of ECs by H_2O_2 and $^{\bullet}OH$ is heavier in high AGE than in high glucose.

Reactive oxygen species is the major cause of injury to ECs, which leads to platelet adhesion in high glucose and high CML-AGE. Other factors still remain to be considered regarding endothelial damage in diabetic conditions.

Conclusion

Enhanced thrombogenesis in diabetes mellitus is mainly caused by endothelial injury due to excess ROS production induced by chronic hyperglycemia and AGEs subsequent to it.

References

1. Tsukada K, Sekizuka E, Oshio C, et al. (2001) Direct measurement of erythrocyte deformability in diabetes mellitus with a transparent microchannel capillary model and high-speed video camera system. Microvasc Res 61(3):231–239
2. Ozawa T, Sekizuka E, Oshio C, et al (2002) Mechanism of increased platelet aggregation in diabetes mellitus—newly-devised investigation with laser scattering platelet aggregometer. Microcirc Annu 18:53–54

T-Lymphocyte Migration is Differently Regulated in Appendiceal Lymph Follicles and Intestinal Peyer's Patches

Yoshikazu Tsuzuki, Hideyasu Nagamatsu, Koji Matsuzaki, Ryota Hokari, Kazuro Itoh, and Soichiro Miura

Key words. T lymphocyte, Appendix, MAdCAM-1, ICAM-1

Appendix as an Inductive Site of Intestinal Inflammation

The appendix was long considered a redundant organ. However, thereafter, analysis of immune components of the appendix revealed that slightly fewer than one third of its cells are T cells and that normal adult appendiceal lymphocyte reactivity is predominated by helper T cells [1]. In addition, the appendix has been reported to be involved in intestinal inflammation. For example, "skip lesions" of the appendix were detected in ulcerative colitis (UC) specimens examined [2], and discontinuous appendiceal involvement was found in patients undergoing proctocolectomy for UC [3]. As regards the mechanisms in appendiceal inflammation, Bittinger et al. demonstrated a different expression of cell adhesion molecules by endothelial (EC) and mesothelial cells (MC) in the various stages of appendicitis, with early E-selectin and intercellular cell adhesion molecule-1 (ICAM-1) expression in EC, followed by vascular cell adhesion molecule (VCAM-1) in EC and MC [4], suggesting that adhesion molecules play pivotal roles in appendiceal inflammation. In addition, recently, potential roles of chemokines and their receptors in lymphocyte migration have been reported.

On the other hand, preventive effects of appendectomy for UC have been reported in a clinical setting [5–8]. Moreover, in an experimental model of young T-cell receptor (TCR)-α deficient mice, removal of the appendix inhibited the induction of experimental colitis [9]. In another experimental model,

Second Department of Internal Medicine, National Defense Medical College, 3-2 Namiki, Tokorozawa, Saitama 359-8513, Japan including the dextran sulfate sodium (DSS)-induced colitis model, appendectomy affords protection against colitis [10].

Although lymph follicles of the appendix (ALFs) play an important role as a secondary lymphoid tissue as described above, the characteristics of lymphocyte trafficking have not been investigated in the appendix. Therefore, we investigated T lymphocyte–endothelial interactions in ALFs compared with those in Peyer's patches (PPs) using an intravital microscope and investigated the contribution of adhesion molecules, mucosal addressin cell adhesion molecule-1 (MAdCAM-1) and ICAM-1, and the role of chemokines in the context of the comparison with that in intestinal PPs [11].

Intravital Observation of Lymphocyte Migration in ALFs

For the preparation of T lymphocytes, spleen and mLNs of BALB/c mice were minced and T-cell-rich fraction was obtained by T-cell separation columns. Thereafter, T lymphocytes were labeled with CFSE solution. The intestine was exteriorized, an appendix and parts of small intestine were ligated, and CFSE-labeled T lymphocytes were injected via a tail vein. Behavior of T lymphocytes was observed from the serosal side by an intravital microscope at three sites (ALF, ileal PP, and jejunal PP) in the same animal and recorded for 90 min. In some sets of experiments, anti-MAdCAM-1 monoclonal antibody (mAb), anti-ICAM-1 mAb, or a control rat IgG was injected 30 min before the infusion of T lymphocytes.

The percentage of rolling lymphocytes in microvessels of ALFs was not significantly different from that in PPs in small intestine. However, the administration of anti-MAdCAM-1 mAb significantly inhibited lymphocyte rolling in ALFs, but not in intestinal PPs. We also found that the number of adherent T lymphocytes was remarkably suppressed by anti-MAdCAM-1 mAb at these three sites. On the other hand, anti-ICAM-1 mAb significantly suppressed T-lymphocyte adherence at 50 min only at the appendix, but not PPs. Finally, chemokine receptor CCR7 was shown to play an important role in T-lymphocyte adherence in all sites. These results suggest the possibility that T-lymphocyte migration is differently regulated in ALFs and intestinal PPs.

References

1. Kawanishi H (1987) Immunocompetence of normal human appendiceal lymphoid cells: in vitro studies. Immunology 60:19–28
2. Kroft SH, Stryker SJ, Rao MS (1994) Appendiceal involvement as a skip lesion in ulcerative colitis. Mod Pathol 7:912–924

3. Perry WB, Opelka FG, Smith D, et al (1999) Discontinuous appendiceal involvement in ulcerative colitis: pathology and clinical correlation. J Gastrointest Surg 3:141–144
4. Bittinger F, Brochhausen C, Kohler H, et al (1998) Differential expression of cell adhesion molecules in inflamed appendix: correlation with clinical stage. J Pathol 186: 422–428
5. Naganuma M, Iizuka B, Torii A, et al (2001) Appendectomy protects against the development of ulcerative colitis and reduces its recurrence: results of a multicenter case-controlled study in Japan. Am J Gastroenterol 96:1123–1126
6. Russel MG, Dorant E, Brummer RJ, et al (1997) Appendectomy and the risk of developing ulcerative colitis or Crohn's disease: results of a large case-control study. South Limburg Inflammatory Bowel Disease Study Group. Gastroenterology 113:377–382
7. Rutgeerts P, D'Haens G, Hiele M, et al (1994) Appendectomy protects against ulcerative colitis. Gastroenterology 106:251–253
8. Smithson JE, Radford Smith G, Jewell GP (1995) Appendectomy and tonsillectomy in patients with inflammatory bowel disease. J Clin Gastroenterol 21:283–286
9. Mizoguchi A, Mizoguchi E, Chiba C, et al (1996) Role of appendix in the development of inflammatory bowel disease in TCR-alpha mutant mice. J Exp Med 184:707–715
10. Krieglstein CF, Cerwinka WH, Laroux FS, et al (2001) Role of appendix and spleen in experimental colitis. J Surg Res 101:166–175
11. Nagamatsu H, Tsuzuki Y, Miyazaki J, et al (2002) Migration of T lymphocytes is differentially regulated in lymph follicles of appendix and Peyer's patches of small intestine. Gastroenterology suppl 122:A153

Cardiotonic Pill Improves Ischemia–Reperfusion-Induced Microcirculatory Disturbances in Rat Mesentery

Jing-Yan Han[1], Yasutada Akiba[1,2], Hidekazu Suzuki[1], Hiroshi Nagata[1], Soichiro Miura[3], and Hiromasa Ishii[1]

Key words. Ischemia–reperfusion, Mesenteric microcirculation, Danshen, Notoginseng, Oxidative stress

Introduction

Cardiotonic Pill (CP), which consists of Danshen and Notoginseng, has clinically been used in China for the medical treatment of vascular diseases, which are related with ischemia–reperfusion (I/R) [1]. Since Danshen, the extract of *Salvia miltiorrhiza*, acts as an oxygen radical scavenger [2], we hypothesized that CP or its gradients, Danshen or Notoginseng, improves the I/R-induced injury as well as the well-known scavengers, superoxide dismutase (SOD) and catalase. We examined the effects of oral administration of CP, Danshen, or Notoginseng on I/R-induced rat mesenteric microcirculatory injury, compared with SOD or catalase.

Materials and Methods

Under sodium pentobarbital anesthesia (50 mg/kg, i.p.), the mesenteric microcirculation of the ileocecal portion of male Wistar rats (200–250 g) was observed through an inverted-type intravital microscope with a CCD or SIT camera as previously described [3]. After 10 min of basal observation, I/R was

[1] Department of Internal Medicine, Keio University School of Medicine, 35 Shinanomachi, Shinjuku-ku, Tokyo 160-8582, Japan
[2] CURE/UCLA & BBRI, Building 114, Suite 217, WLA VAMC, 11301 Wilshire Boulevard, Los Angeles, CA 90073, USA
[3] Second Department of Internal Medicine, National Defense Medical College, 3-2 Namiki, Tokorozawa, Saitama 359-8513, Japan

conducted by a ligation of the mesenteric artery and vein for 10 min followed by the release of the blood flow occlusion. We measured the venular diameter, the number of adherent leukocyte, fluorescein isothiocyanate-albumin leakage as vascular permeability, dihydrorhodamine-123 (DHR) fluorescence to monitor oxidative stress related with hydrogen peroxide, and mast cell degranulation [3], with or without the administration of CP, Danshen, or Notoginseng (each 0.4 g/kg) via an orogastric tube 1 h before I/R. Superoxide dismutase (12 000 units/kg per hour) or catalase (20 mg/kg per hour) was continuously injected i.v. from 20 min before I/R. In some rats, the observed area of mesentery was removed and fixed with 1% paraformaldehyde solution for immunofluorescence using anti-ICAM-1 or anti-CD11b antibody to examine the expression of the adhesion molecules after I/R.

Results

Ischemia–reperfusion increased the number of adherent leukocytes, albumin leakage, DHR fluorescence, and mast cell degranulation. Cardiotonic Pill as well as SOD attenuated these changes observed after I/R, whereas catalase only reduced the number of adherent leukocytes and DHR fluorescence. Danshen reduced the number of adherent leukocytes, albumin leakage, and DHR fluorescence but not mast cell degranulation, whereas Notoginseng reduced the number of adherent leukocytes and mast cell degranulation, but not albumin leakage or DHR fluorescence (Fig. 1). Immunofluorescence study showed an abundant ICAM-1 expression in the endothelium with CD11b-positive leukocytes in the I/R alone group, whereas CP decreased the endothelial ICAM-1 expression and CD11b-positive leukocytes.

Discussion

We demonstrated that the oral administration of CP diminished the microcirculatory alteration induced by I/R as well as SOD and rather than catalase, suggesting that CP has an antioxidative effect on both O_2^- and H_2O_2, consistent with the scavenger effect of Danshen on the reactive oxygen species [2]. Cardiotonic Pill also inhibited the leukocyte adhesion to the endothelium accompanied with the reduced expression of ICAM-1 and CD11b, presumably due to its effect on the cell membrane metabolism, since the extract of *S. miltiorrhiza* inhibits the arachidonic acid metabolism and reduces the inflammatory responses [4]. We also demonstrated that the effects of CP are due to the combination of Danshen and Notoginseng, each having different effective sites on the microcirculatory parameters. Our study suggests that orally administered CP protects the endothelium through the inhibition of

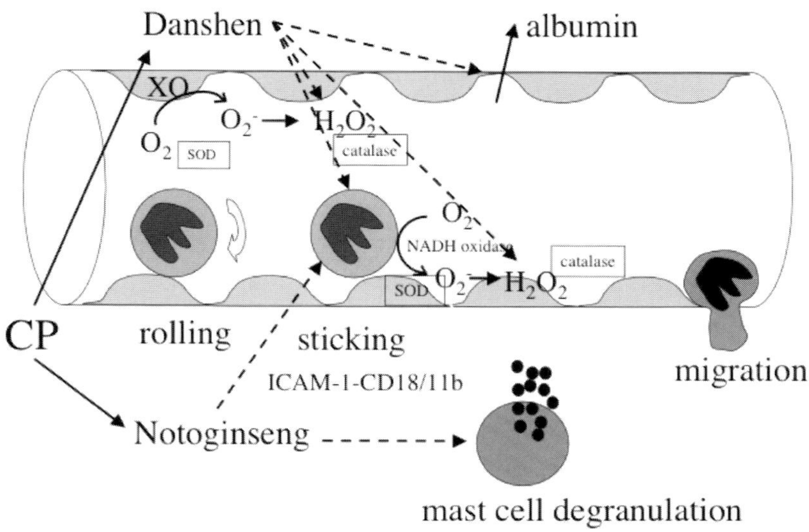

FIG. 1. Effects of Cardiotonic Pill (*CP*) and its gradients, *Danshen* and *Notoginseng*, on the microcirculatory disturbances induced by ischemia–reperfusion in rat mesentery. Danshen, one of the CP gradients, inhibits leukocyte adhesion, H_2O_2 production, and albumin leakage, whereas the other gradient, Notoginseng, inhibits leukocyte adhesion and mast cell degranulation. *Dashed lines* show the inhibitory effects. *XO*, xanthine oxidase; *SOD*, superoxide dismutase

leukocyte–endothelium interaction and the reduction of oxidative stress from I/R-induced injury.

References

1. Ji XY, Tan BK, Zhu YZ (2000) *Salvia miltiorrhiza* and ischemic diseases. Acta Pharmacol Sin 21:1089–1094
2. Han JY, Miura S, Akiba Y, et al (2001) Chronic ethanol consumption exacerbates microcirculatory damage in rat mesentery after reperfusion. Am J Physiol 280:G939–G948
3. Kang HS, Chung HY, Byun DS, et al (2003). Further isolation of antioxidative (+)-1-hydroxypinoresinol-1-O-beta-D-glucoside from the rhizome of *Salvia miltiorrhiza* that acts on peroxynitrite, total ROS and 1,1-diphenyl-2-picrylhydrazyl radical. Arch Pharm Res 26:24–27
4. Kim SY, Moon TC, Chang HW, et al (2002) Effects of tanshinone I isolated from *Salvia miltiorrhiza* Bunge on arachidonic acid metabolism and in vivo inflammatory responses. Phytother Res 16:616–620

Molecular Determinants of Dextran Sodium Sulfate Colitis-Induced Platelet Adhesion in Colonic Venules

Mikiji Mori, James W. Salter, Thorsten Vowinkel, and D. Neil Granger

Key words. Inflammatory bowel disease, Animal model, Microcirculation, Selectin, Integrin

Introduction

Platelets have been implicated in the pathogenesis of human inflammatory bowel diseases (IBD). Animal experimentation has recently focused on defining the magnitude and mechanism of platelet accumulation in the inflamed bowel. We have previously shown that the recruitment of adherent platelets in colonic venules of mice with dextran sodium sulfate (DSS)-induced colitis can be blocked by a P-selectin monoclonal antibody (mAb) and is attenuated in P-selectin deficient mice. The objectives of this study were to determine whether: (1) the counter-receptor for P-selectin mediated recruitment of platelets in the DSS colitis model is PSGL-1, and (2) immunoblockade of glycoprotein IIb/IIIa (GPIIb/IIIa) affects platelet recruitment in DSS colitis.

Materials and Methods

Colitis was induced by feeding wild type (WT) C57Bl/6 mice 3% DSS in drinking water for 6 days. The DSS treatment protocol resulted in weight loss, bloody stools, and colonic mucosal injury. The estimate of inflammation in DSS-induced colitis was expressed as disease activity index (DAI) [1]. Intravital microscopy was used to observe and quantify the adhesion of platelets in colonic venules. Platelets isolated from WT mice were labeled ex vivo with the

Department of Molecular and Cellular Physiology, Louisiana State University Health Sciences Center, 1501 Kings Highway, Shreveport, LA 71130-3932, USA fluorochrome carboxyfluorescein diacetate succinimidyl ester (CFSE) and then platelets (10×10^7) were injected intravenously into either control or colitic mice. Some colitic mice were pretreated intravenously with a mAb directed against either PSGL-1 or GPIIb/IIIa (2 mg/kg) 30 min before obtaining adhesion measurements.

Results

The disease activity index score was elevated significantly from day 4 in DSS-treated mice. The venules in colitic mice exhibited a profound increase in the number of adherent platelets compared with control mice. The DSS-induced recruitment of adherent platelets was markedly attenuated in WT mice receiving a PSGL-1 mAb. However, pretreatment with a GPIIb/IIIa mAb did not alter DSS-induced platelet adherence (Fig. 1).

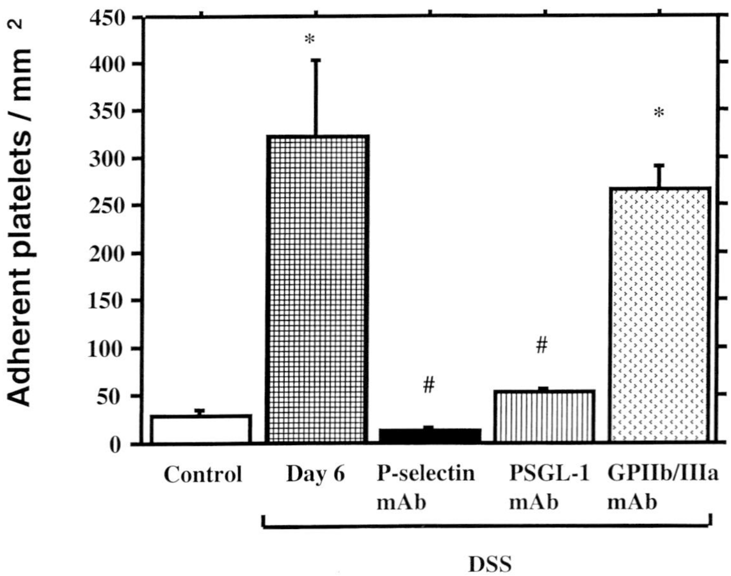

FIG. 1. Effects of P-selectin monoclonal antibody (mAb), PSGL-1 mAb, and glycoprotein IIb/IIIa (GPIIb/IIIa) mAb in dextran sodium sulfate (DSS)-induced platelet adherence. Dextran sodium sulfate-induced colitic venule had a 12-fold increment in the number of adherent platelets compared with control. The recruitment of the platelets was abrogated significantly in DSS-mediated colitic mice after treatment with P-selectin mAb, and pretreatment with PSGL-1 mAb also prevented DSS-induced adherent platelet. However, pretreatment with a GPIIb/IIIa mAb did not cause any alteration. $^*P < 0.05$ vs control, $^\#P < 0.05$ vs DSS on day 6

Discussion

Our data suggest that platelet-associated P-selectin is the major mediator of DSS-induced platelet adherence and that PSGL-1 is a ligand for P-selectin, based on the reduction in the number of adherent platelets observed in colitic mice treated with the PSGL-1 mAb. This suggests that PSGL-1 expressed on endothelial cells or on leukocytes already adherent to endothelial cells engages with P-selectin expressed on the surface of activated platelets. There is evidence already in the literature that supports an increased expression of P-selectin on circulating platelets in patients with IBD [2]. We also tested the possibility that the β_3-integrin GPIIb/IIIa contributes to the platelet adhesion associated with colitis. Using a GPIIb/IIIa blocking mAb, we found no evidence for a role for GPIIb/IIIa in this model of inflammation. In conclusion, these findings support a role for P-selectin (but not GPIIb/IIIa) in mediating the platelet recruitment in inflamed colonic venules and implicate PSGL-1 as the ligand for these adhesive interactions.

References

1. Cooper HS, Murthy SN, Shah RS, et al (1993) Clinicopathologic study of dextrin sulfate sodium experimental murine colitis. Lab Invest 69:238–249
2. Schurmann GM, Bishop AE, Facer P, et al (1995) Increased expression of cell adhesion molecule P-selectin in active inflammatory bowel disease. Gut 36:411–418

Lymphatic Vessels as a Selective Administration Route of Antitumor Agent

Akira Fujimura[1], Masao Onodera[1], Nobuo Saito[2], Toshihiro Ohta[2], Kohichiro Sera[3], and Yohichiro Nozoka[1]

Key words. Lymphatic vessel, Drug delivery route, Selective administration route, Tumor, Antitumor agent

Introduction

Enucleation of primary lesion and neck dissection of regional lymph nodes are required in operating on a tumor. As a result, rehabilitation is very difficult following treatment of a tumor, because of the large defect and the side effects. The dissection of regional lymph nodes is required in the operation of mammary cancer and uterine cancer, causing edema of the limb. In other words the quality of life of a patient is adversely affected by current surgical methods. This has been improved by the use of Sentinel navigation surgery, but it is reported that this method is impossible to use in the head and neck region due to the existence of the lymphatic network. We surmised that the quality of life of the patient could be maintained by using the lymph vessels as a drug delivery route after involving the lymphatic network. From our past reports [1–4], we are able to elucidate, to a great extent, the detailed architecture of lymphatic vessels in the tongue. However, the network of the lymph nodes in the neck region is very complicated and at present is far from being elucidated. On the other hand, the lymph vessels from every organ connect with the regional lymph nodes as stated in the theory of Sentinel navigation

[1] First Department of Oral Anatomy, [2] First Department of Maxillofacial Surgery, [3] Cyclotron Research Center, Iwate Medical University, 1-2-27 Chuo-dori, Morioka 020-8505, Japan

surgery and as discovered by research on the lymph vessels. The lymph node dissection which is done currently intercepts the lymph flow from other organs. We propose a method which does not dissect the regional lymph nodes where metastasis is clear. There have been some reports that propose the intratumoral dosage of antitumor agents, but as yet there are no reports regarding use of the lymph vessels as a drug delivery route to metastatic lymph nodes selectively with a high concentration of agent. We injected an antitumor agent into the normal tongue as a preliminary study and confirmed the flow to regional lymph nodes.

Materials and Methods

Mice were injected with 5 and 10 μl of cisplatin (1 mg/ml) (MW: 300.05, Randa, Nippon Kayaku, Japan) to the right side of the normal tongue and taking in the right and left submandibular lymph nodes, blood, liver, and kidney, 0.5, 1, 3, 6, and 24 h after injection of cisplatin. Samples were dried and prepared by the chemical ashing method using nitric acid, and then analyzed by particle induced X-ray emission. Cisplatin includes platinum which does not exist in generally in organisms, and this free element aids antitumor activity. Therefore, we can estimate the antitumor activity by measuring the dose of platinum.

Results and Discussion

Platinum was scarcely detected within 24 h in the 5 μl injection group, and only in the submandibular lymph nodes which was the regional lymph node in the 10 μl injection group (Fig. 1). It is clear that the pressure in the tissue triggered absorption by the initial lymph vessels. Cisplatin flowed into both sides of the submandibular lymph nodes and gradually increased proportionately with time. It was clear that the cisplatin had a high concentration locally, but a low concentration in the whole body. We believe that by use of this method using the lymphatic vessels as a drug delivery route, the burden on the patient during operation will be reduced and the side effects of the antitumor agent will be inhibited.

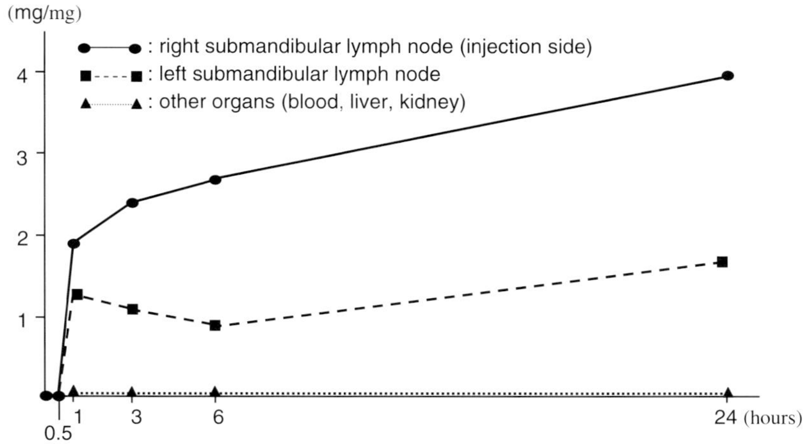

FIG. 1. Time course analysis of platinum by particle induced X-ray emission. Platinum increases throughout the time course. Regarding quantity of platinum, the submandibular lymph node of the injection side contains less than half that of the one on the opposite side. Platinum could not be detected in other organs

References

1. Xie X, Fujimura A (2000) Macromolecular clearance in the tongue through lymphatic vessels. Dent J Iwate Med Univ 25:273–282
2. Fujimura A, Nozaka Y (2002) Analysis of the three-dimensional lymphatic architecture of the periodontal tissue using a new 3D reconstruction method. Microsc Res Tech 56:60–65
3. Fujimura A, Seki S, Liao M-Y, et al (2003) Three dimensional architecture of lymphatic vessels in the tongue. Lymphology 36:120–127
4. Seki S, Fujimura A (2003) Three dimensional changes in lymphatic architecture around VX2 tongue cancer. Dynamics of growth of cancer. Lymphology 36:128–139
5. Seki S, Fujimura A (2003) Three dimensional changes in lymphatic architecture around VX2 tongue cancer. Dynamic changes after administration of antiangiogenic agent. Lymphology 36:199–208

Part 5
Gaseous Molecules: Sensing and Bioregulation in Microcirculation

The Heme Oxygenase–Carbon Monoxide System as a Regulator of Microvascular Function

Makoto Suematsu

Summary. Heme oxygenase (HO) catalyzes oxidative cleavage of protoheme IX to generate divalent iron, biliverdin, and carbon monoxide (CO). The interest in the HO–CO system has emerged in numerous disciplines among such as cardiovascular physiology, the central nervous and hepatic microvascular systems. Although for many years products of the HO reaction had been regarded as potentially toxic wastes, recent studies have implicated that these products play physiological roles. Both NO and CO share the ability to bind to the prosthetic group of heme proteins, structural changes and the functional outcomes of the proteins seem quite different between the gases. Differences in effects on soluble guanylate cyclase and hemoglobin between NO and CO led us to understand mechanisms as to how the proteins can distinguish the gases to transducer signals in distinct ways. This chapter focuses on recent advances in both physiologic and pathophysiologic roles of CO and aims to provide updated information on these gas mediators as potential regulators of the organ function on the basis of data collected from the model of isolated perfused liver of rats.

Key words. Heme oxygenase, Carbon monoxide, Guanylate cyclase, Heme protein, Cytochrome P450

Introduction

Heme oxygenase (HO) degrades protoheme IX, giving rise to carbon monoxide (CO), ferrous iron, and biliverdin, which is rapidly reduced to bilirubin [1]. Amongst the three products of HO, CO has been extensively studied as a

Department of Biochemistry and Integrative Medical Biology, Keio University School of Medicine, 35 Shinanomachi, Shinjuku-ku, Tokyo 160-8582, Japan potential neurotransmitter in the brain [2,3] and as a gaseous mediator in the liver [4,5]. The products of HO reaction including CO and bilirubin appear to modulate a variety of hepatobiliary functions in the liver. CO has been found to be the most important regulator of sinusoidal blood flow in normal liver [4]. This is based on our observation that elimination of endogenous CO by zinc protoporphyrin IX (ZnPP), a potent inhibitor of HO, caused to increase in the vascular resistance in the isolated perfused liver. The mechanisms for the gas reception and sinusoidal relaxation appear to involve soluble guanylate cyclase (sGC).

While CO shares the ability to activate sGC with NO, the two gases turned out to possess quite different properties to alter the function of heme proteins. We herein intend to compare and contrast roles of CO with those of NO, another gaseous monoxide that functions as a signal in diverse physiological processes. The parallelism exists between the two gaseous molecules as such the two monoxides are notable among signals for their rapid diffusion and ability to permeate cell membranes. Although CO and NO are structurally similar, there is a distinctive property in their chemical reactivity to the prosthetic heme of the proteins. Considering the similarity and the difference, the present chapter attempts to address the possible mechanisms for mutual regulatory interactions between CO and NO.

Microvascular Actions of CO: Why Not NO in the Liver?

In mammals, HO exists in two forms: HO-1 and HO-2. HO-1 is induced by varied stressors such as cytokines, heavy metals, ROS and hypoxia. Excess NO could also cause the HO-1 induction. Microvascular actions of endogenously generated CO was first demonstrated in the liver [4–6]. We demonstrated intrahepatic distribution of two major HO isozymes immunohistochemically, with the finding that the two isozymes have distinct topographic patterns; HO-1, the inducible form, is expressed prominently in Kupffer cells, while the constitutive HO-2 is abundant in hepatocytes [6]. Carbon monoxide derived from HO-2 is necessary to keep sinusoids in a relaxing state through mechanisms involving sGC in hepatic stellate cells (HSC), also known as Ito cells that constitute microvascular pericytes in this organ. Considering the microanatomical orientation of the liver cells in and around sinusoids, HO-2 in parenchyma stands in the reasonable position for the gas reception by HSC where CO released from hepatocytes can directly access to the cells and thereby modulate their contractility without being captured by hemoglobin in circulation. When exposed to disease conditions such as endotoxemia and advanced cirrhosis, liver could upregulate HO-1 in Kupffer cells and hepatocytes as a result of cytokine responses [7]. In experimental models of endotoxemia, such an induction of HO-1 expands the ability of liver to degrade heme and to trigger overproduction of CO. Under these circumstances, CO turned out to contribute to maintenance of blood perfusion as well as that of bile excretion.

As mentioned, HO-1 is an inducible enzyme, and with exposure to stress stimuli such as cytokine, hypoxia and superoxide, HO-1 increases not only in Kupffer cells but also in parenchymal cells [6,7]. It should be noted that under the normal condition, nitric oxide synthase (NOS) is expressed in the sinusoidal endothelium; however, under the pathophysiological condition, an inducible form of NOS can be expressed in hepatocytes. This lack of NO production in the hepatocytes in normal liver may be of a physiological importance in some ways. The first importance may arise when we consider the fact that the hepatocyte is the locus of the urea cycle where arginine (a substrate of NOS) and citrulline (by-product of NOS reaction) are the playing partners. Under the circumstance that hepatocytes start expressing inducible NOS in response to cytokine, the efficiency of urea cycle may be reduced due to a possible shunt between arginine and citrulline created by catalytic action of inducible NOS. The shunt of this kind is undesirable for the efficient elimination of ammonia. In addition, it has been reported that NO binds to the heme moiety of cytochrome c oxidase, and consequently it causes mitochondrial dysfunction. This could represent the second importance of not having much NO in hepatocytes.

The role of CO in sinusoidal relaxation was clearly shown by experiments using varied forms of hemoglobin; namely oxyhemoglobin (HbO_2, a ferroheme compound that traps both CO and NO) and methemoglobin (metHb, a ferriheme form that traps NO alone but not CO). Of the two, only the HbO_2 was able to reproduce the vasoconstrictor effect of ZnPP, a potent HO inhibitor [4–6]. Furthermore, oxyhemoglobin that was encapsulated in SIM!250 nm-diameter liposome (HBV-O_2) so that it was restricted to the sinusoids, failed to induce vasoconstriction, suggesting that the locus of action of CO is extrasinusoidal [6]. Considering the fact that free hemoglobin originated from senescent erythrocytes is immediately oxidized to metHb and is metabolized either in Kupffer cells or in the hepatocytes, one might realize a problematic design if NO were the dominant mediator to relax vascular tone. If so, the delivered metHb would bind to NO before it could act on a receptor protein(s), most likely to soluble guanylate cyclase, and this elimination of NO would make it impossible to maintain low vascular resistance in the hepatic microcirculation. In other words, using CO not NO appears to be a clever design to maintain low vascular tone in the hepatic microcirculation under physiological conditions.

Carbon Monoxide Effects on Heme Proteins: Analogy and Difference with NO

Both NO and CO share the binding ability to heme proteins, functional outcome occurring on the proteins seems quite different between the two gases. In case of hemoglobin, CO stabilizes the six-coordinated form of the prosthetic heme and increases the affinity of molecular oxygen in other subunits, whereas NO binds to the α subunit of the heme and breaks the proximal histidine-Fe bond, forming a five-coordinated nitrosyl heme complex to decrease the affinity of oxygen in β subunits. Likewise the case of hemoglobin, differences between NO and CO in the heme structure in the β subunit of sGC appear to cause distinct activation states of the catalytic α subunit of the enzyme. Because of such a structural difference in the heme coordination between NO and CO, the interaction of the two gases on the prosthetic heme of the enzyme leads to a unique regulatory response of the enzyme: low tissue NO makes CO a modestly stimulatory modulator of the enzyme, whereas high tissue NO makes CO an inhibitory one. Observation that vascular smooth muscle cell-specific heme oxygenase-1 transgenic mice exhibit systemic hypertension rather than hypotension supports such a possibility [8]. This notion was also confirmed by our recent studies by showing that the interactions between the two gases cause fine-tuning of the sGC function in vivo [9]. In this study, we applied the newly developed monoclonal antibody (mAb) 3221 against sGC that can recognize the specific structure produced by the enzyme activation. Immunohistochemical analyses of rat retina where the background NO-generating activities appear heterogeneous among different neuronal layers revealed that light-induced upregulation of HO-1 activates sGC in retinal pigment epithelia (low NO), while suppressing the enzyme in inner plexiform layer (high NO). Physiologic roles of CO in this particular organ have not fully been investigated. However, distinct from NO, retina could benefit from the non-radical CO to maintain housekeeping cyclic guanosine monophosphate without a risk of potential degradation of retinoids. The detail description on difference between NO- and CO-mediated signaling events is also available in another review article published from our laboratory [10]. Such a way to use CO is likely to be the case in relaxation of hepatic stellate cells to guarantee sinusoidal patency or in apoptotic control of spermatogenesis, where NO-breakable DNA or vitamin A is abundantly stored, respectively [10].

Acknowledgments. The authors acknowledge support by the 21st Century Center-of-Exellence Program and by the Leading Project for Biosimulation from the Ministry of Education, Sciences, and Technology of Japan. A portion of the Project was also supported by a Grant-in-Aid for Creative Scientific Research by the Japan Society for the Promotion of Sciences 16GS0015.

References

1. Maines MD, Kappas A (1974) Cobalt induction of hepatic heme oxygenase; with evidence that cytochrome P-450 is not essential for this enzyme activity. Proc Natl Acad Sci USA 71:4293–4297
2. Verma A, Hirsch DJ, Glatt CE, et al (1993) Carbon monoxide: a putative neural messenger. Science 259:381–384
3. Poss KD, Thomas MJ, Ebralidze AK, et al (1995) Hippocampal long-term potentiation is normal in heme oxygenase-2 mutant mice. Neuron 15:867–873
4. Suematsu M, Kashiwagi S, Sano T, et al (1994) Carbon monoxide as an endogenous modulator of hepatic vascular perfusion. Biochem Biophys Res Commun 205:1333–1337
5. Suematsu M, Goda N, Sano T, et al (1995) Carbon monoxide: an endogenous modulator of sinusoidal tone in the perfused rat liver. J Clin Invest 96:2431–2437
6. Goda N, Suzuki K, Naito M, et al (1998) Distribution of heme oxygenase isoforms in rat liver: Topographic basis for carbon monoxide-mediated microvascular relaxation. J Clin Invest 101:604–612
7. Kyokane T, Norimizu S, Taniai H, et al (2001) Carbon monoxide from heme catabolism protects against hepatobiliary dysfunction in endotoxin-treated rat liver. Gastroenterology 120:1227–1240
8. Imai T, Morita T, Shindo T, et al (2001) Vascular smooth muscle cell-directed overexpression of heme oxygenase-1 elevates blood pressure through attenuation of nitric oxide-induced vasodilation in mice. Circ Res 89:55–62
9. Kajimura M, Shimoyama M, Tsuyama S, et al (2003) Visualization of gaseous monoxide reception by soluble guanylate cyclase in rat retina. FASEB J 17:506–508
10. Kajimura M, Goda N, Suematsu M (2002) Organ design for generation and reception of CO: Lessons from the liver. Antioxid Redox Signal 4:633–637

Biological Characteristics of Lymphatic Endothelial Cells with Special Reference to Nitric Oxide and Lymphangiogenesis

Toshio Ohhashi

Summary. Biological properties of lymphatic endothelial cells including measurement of partial pressure of oxygen, establishment of rat lymphatic endothelial cell line, immunoreactivity of endothelial constitutive nitric oxide (NO) synthase in the cultured lymphatic endothelial cells, NO-mediated modulation of spontaneous contractions in isolated lymph vessels and lymph pump activity in vivo, and flow-mediated release of NO from lymphatic endothelial cells were shown with our current studies. By using the cultured lymphatic endothelial cells, basic fibroblast growth factor-mediated lymphangiogenesis in vitro is also demonstrated.

Key words. Po_2 in lymph, Lymphatic endothelial cell line, Nitric oxide, Lymphangiogenesis, Basic fibroblast growth factor

Introduction

The lymphatic system returns fluid and protein to the circulation by mechanisms that are not completely understood. How are fluid and protein transported from the tissue spaces into the lymph capillaries (lymph formation)? Many new theories are being developed but the problem is complicated by our uncertainty about tissue fluid pressure. Once lymph is formed, how is it transported back to the general circulation? Some lymph vessels are known to contract intrinsically, and the contraction may play a significant role in the centripetal propulsion of lymph [1–3]. We have studied the physiology and pathophysiology (lymph edema and lymphatic metastasis of carcinoma cells) of lymph formation and lymphatic transport. This chapter summarizes our

Department of Physiology, Shinshu University School of Medicine, 3-1-1 Asahi, Matsumoto 390-8621, Japan findings in four areas: (1) the partial oxygen pressure of lymph in vivo, (2) establishment of rat lymphatic endothelial cell line, (3) functional roles of nitric oxide (NO) in lymphatic transport, and (4) basic fibroblast growth factor (bFGF)-mediated lymphangiogenesis in vitro.

Partial Pressure of Oxygen in Lymph In Vivo

There exists a discrepancy regarding on partial pressure of oxygen (Po_2) of lymph; some studies showed the values lower than the Po_2 in venous blood and the others the values similar to the value in venous blood [4]. Thus we measured changes in the Po_2 of lymph in thoracic ducts in vivo by using oxygen electrodes and then examined the effects of 3 mM KCl on changes in the Po_2 of lymph and lymph flow rate. Mongrel dogs were anesthetized with sodium pentobarbital (30 mg/kg, i.v.) and artificially ventilated with room air by a respirator. A catheter was inserted into the right femoral vein to administrate physiological saline solution or the high-potassium solution and another one into the right femoral artery to monitor changes in the systemic arterial pressure. A polyethylene catheter connected with a glass capillary and oxygen electrode was inserted into the intrathoracic duct from the left jugular angle. The partial pressure of oxygen in lymph was continuously measured with an oxygen electrode and its electrical amplifier. The flow rate of lymph was also recorded by using a home-made drop-counter. The Po_2 values of arterial and venous blood were 110.5 ± 8.0 mmHg and 55.2 ± 3.5 mmHg, respectively ($n = 7$). As shown in Fig. 1, the Po_2 value of lymph in the thoracic duct was around 35 mmHg in vivo.

The cardiac arrest produced by an intravenous administration of 3 mM KCl caused a rapid decrease of the arterial pressure that became around 0 mmHg by 1 min after the cardiac arrest. The cardiac arrest also produced a gradual decrease of the Po_2 value that arrived at a stable value (10 mmHg) by 15 min after the cardiac arrest. In contrast the lymph flow rate was rapidly and magnificently increased by the cardiac arrest, the slight increase of which was significantly kept during 15 min after the cardiac arrest [5]. These findings suggest that lymphatic system work physiologically under 35–40 mmHg in Po_2.

Establishment of Rat Lymphatic Endothelial Cell Line

The cultured lymphatic endothelial cells of dogs and cows have been used to investigate biological and morphological properties of the endothelial cells [6,7]. No report, however, exists regarding the establishment of lymphatic endothelial cell line from small experimental animals such as rats and mice.

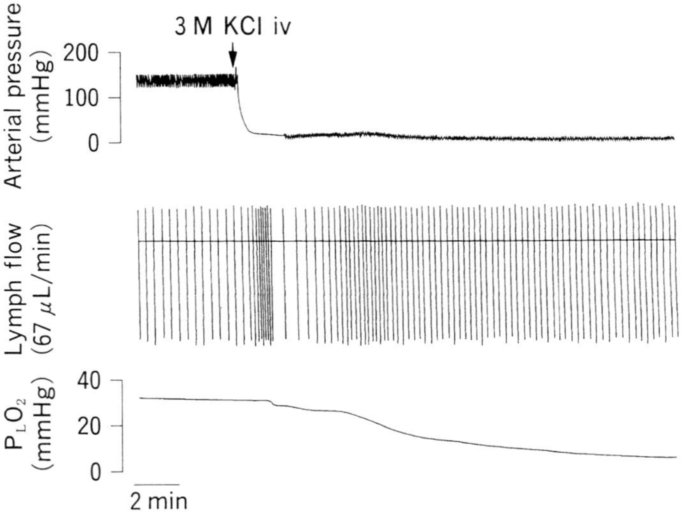

FIG. 1. Representative tracings of effects of 3 mM KCl-mediated cardiac arrest on arterial pressure (*upper panel*), lymph flow rate of thoracic duct recorded with a home-made drop counter (*middle panel*), and partial pressure of oxygen in the intrathoracic lymph measured with an oxygen electrode in an anesthetized dog (*lower panel*)

The establishment may facilitate the promotion of key studies to evaluate cellular and molecular mechanisms of lymphangiogenesis. Thus we have attempted to establish rat lymphatic endothelial cell line and then to investigate morphological and immunohistochemical properties of the cultured cells and arrangement of cytoskeleton protein F-actin [8]. The lymphatic endothelial cells of rat thoracic ducts were isolated enzymatically by trypsin digestion and were cultured in endothelium growth medium (EGM)-2 in an atmosphere of low oxygen (5% O_2, 5% CO_2, and 90% N_2) or high oxygen (21% O_2, 5% CO_2, and 74% N_2). The number of the cells cultured in the low-oxygen atmosphere (48 750 ± 10 594 cells/ml, $n = 4$) was significantly larger than that obtained in the high-oxygen atmosphere (4333 ± 1377 cells/ml, $n = 4$). The cultured cells in the low-oxygen atmosphere showed a monolayer with uniform cobblestone appearance, suggesting the morphological properties of endothelial cells (Fig. 2). Factor VIII-related antigen and cell surface carbohydrates were found on the lymphatic cultured cells. The phagocytosis of 1,1-diocadecyl1-3,3,3′,3′-tetramethylindo-carbocyanine perchlorate-labeled acetylated low-density lipoprotein also was observed in the cultured cells. The cytoskeleton protein F-actin was located on the plasma membrane of the cultured cells as circumferential thin bundles and in the cytoplasm as filamentous bundles. The study indicates that the choice of EGM-2 as a culture

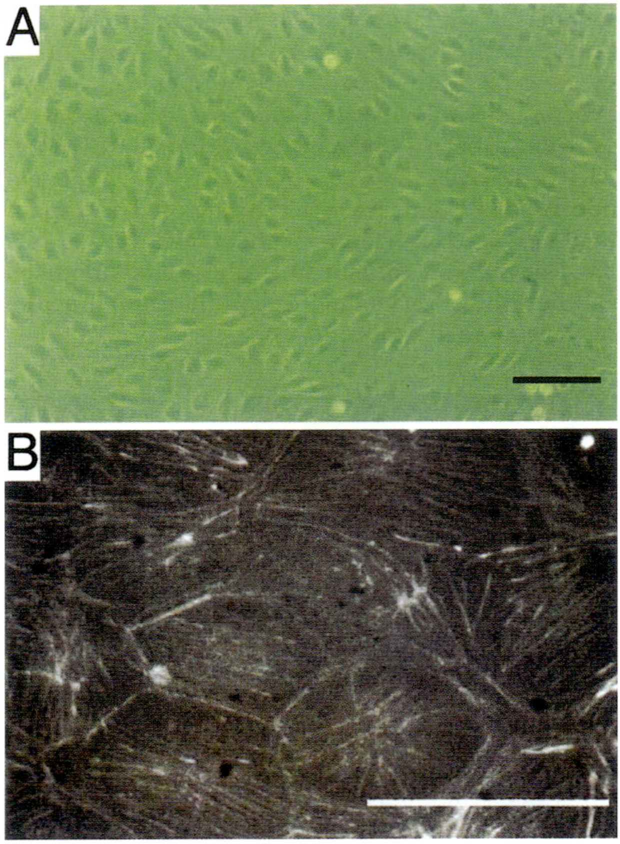

Fig. 2. Representative photographs of a phase-contrast image (**A**) and of the arrangement of F-actin (**B**) in cultured rat lymphatic endothelial cells. The bars in **A** and **B** are 100 and 50 μm, respectively. From ref. [8]

medium and the hypoxic atmosphere (~5%) enabled us to establish rat lymphatic endothelial cell line.

Physiological Roles of Endogenous NO in Lymphatic System

The immunoreaction to antiendothelial constitutive NO synthase (ecNOS) was significantly positive to the cultured lymphatic endothelial cells (LEC) [9]. When we stained 14 samples of the LEC to the anti-ecNOS, the immunoreactive signals were intense in the nucleus and cytoplasm (10 out of 14). In

4 of 14 samples, the intense signal of anti-ecNOS was restricted in the nuclei.

The NO released from lymphatic endothelial cells can regulate the rhythm and amplitude of spontaneous contractions in isolated bovine mesenteric lymph vessels [10]. Regular spontaneous contractions at a constant rate of about 3 beats/min were observed in the isolated lymph vessels. Acetylcholine (ACh) at concentrations between 10^{-7} and 10^{-6} M caused both negative chronotropic and inotropic effects on the spontaneous contractions. The ACh-mediated negative effects were completely reversed in all lymphatic segments studied when the endothelium was removed mechanically. Addition of 3×10^{-5} M N^{G}-monomethyl-L-arginine (L-NMMA) tended to increase the rhythm and amplitude of the spontaneous contractions in the control. The ACh-mediated negative chronotropic and inotropic effects in the lymphatic segments with intact endothelium were significantly reduced by the pretreatment with L-NMMA. An additional treatment with 10^{-4} M L-arginine in the same segments caused a complete reversal of the ACh-mediated chronotropic and inotropic effects on the spontaneous contractions. The findings suggest that NO liberated from the lymphatic endothelium seems to inhibit pacemaker activity of the spontaneous contractions and to reduce myogenic conduction and/or the mechanical activity of the lymphatic smooth muscles.

We also examined effects of flow (shear stress) on lymphatic endothelial cells by using conventional cascade bioassay preparations [11]. The pressurized canine thoracic ducts were intraluminally circulated at flow rate ranging from 0.5 to 2.0 ml/min. A linear relationship between the flow rate and the normalized amount of NO released from the lymphatic endothelial cells was observed during the range of flow rate. Thus the lymphatic endothelial cells are very sensitive to lowest changes of shear stress compared with arterial and venous endothelial cells.

Next we attempted to examine effects of N^{ω}-nitro-L-arginine methyl ester (L-NAME) on the pump activity of rat mesenteric lymph vessels in vivo by using a vital video microscope [12]. The exposed surface of rat mesentery was continuously perfused with 37°C bicarbonate-buffered physiological salt solution. Pumping frequency (PF), end-diastolic (EDD) and end-systolic diameters (ESD) of the mesenteric lymph microvessels were measured with the vital microscopic system and then the pump flow index (PFI) was calculated. A 15-min perfusion of 30 μM L-NAME over the mesenteries caused a significant increase of the PF and PFI and a significant decrease of the EDD and ESD. Simultaneous perfusion of 1 mM L-arginine with 30 μM L-NAME produced a significant reversal of the L-NAME-mediated increase of PF and decrease of ESD. The findings suggest that endogenous NO, may be released by the stimulation of flow, have physiologically modulated the pump activity in rat mesenteric lymph vessels in vivo.

Basic Fibroblast Growth Factor-Mediated Lymphangiogenesis In Vitro

Elevated interstitial fluid pressure in human tumors has been attributed to the increased permeability of tumor vessels, the growth of the vessels in a confined space, and the absence of a well-defined lymphatic system [13]. We examined whether the cultured lymphatic endothelial cells can induce in vitro neovascularization of lymph vessels, similar to angiogenesis of the blood vessels, in response to basic fibroblast growth factor (bFGF) [14]. The effects of bFGF on the proliferation and migration of cultured lymphatic endothelial cells were evaluated by changing the number of the subconfluent cells and by wound migration assay, respectively. We also examined effects of bFGF on invasion of the cultured lymphatic endothelial cells into a three-dimensional collagen gel by using a phase-contrast microscope (Fig. 3) and an electron microscope. The bFGF caused significant invasion and tube formation into the three-dimensional collagen gel. The growth factor also facilitated formation of capillary-like-tubes of the cultured cells between two layers of colla-

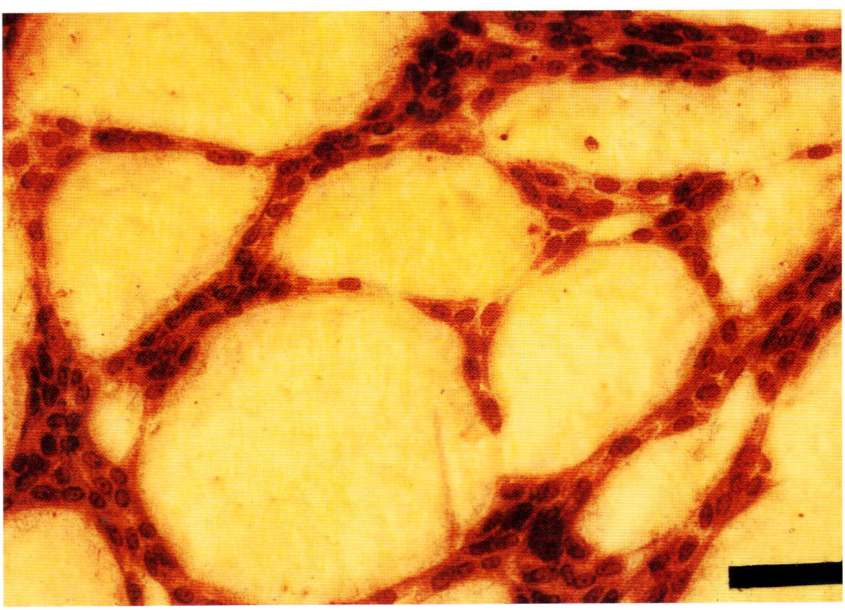

Fig. 3. A representative microphotograph demonstrating tube formation (lymphangiogenesis) of the cultured lymphatic endothelial cells when the cells were sandwiched between two layers of type I collagen gels in the presence of 10 ng/ml basic fibroblast growth factor. Bar = 100 μm

gen gels. The findings suggest that the cultured lymphatic endothelial cells can form lymphatic capillary-like tubes in response to bFGF.

References

1. Hall JG, Morris B, Woolly G (1965) Intrinsic rhythmic propulsion of lymph in the unanesthetized sheep. J Physiol (Lond) 180:336–349
2. Mawhinney HIJ, Roddie IC (1973) Spontaneous activity in isolated bovine mesenteric lymphatics. J Physiol (Lond) 229:339–348
3. Ohhashi T, Azuma T, Sakaguchi M (1980) Active and passive mechanical characteristics of bovine mesenteric lymphatics. Am J Physiol 239:H88–H95
4. Bergofsky EH, Jacobson JH, Fishman AP (1962) The use of lymph for the measurement of gas tension in interstitial fluid and tissues. J Clin Invest 41:1971–1980
5. Ikomi F, Mizuno R, Ohhashi T (2000) Effects of vasoactive substances on oxygen tension of lymph in thoracic duct. Jpn J Physiol 50(suppl):S74
6. Leak LV, Jones M (1993) Lymphatic endothelium isolation, characterization and long term culture. Anat Rec 236:641–652
7. Jones BE, Young LCJ (1987) Culture and characterization of bovine mesenteric lymphatic endothelium. In Vitro Cell Dev Biol 23:698–706
8. Mizuno R, Yokoyama Y, Ono N, et al. (2003) Establishment of rat lymphatic endothelial cell line. Microcirculation 10:127–131
9. Nojiri H, Ohhashi T (1999) Immunolocalization of nitric oxide synthase and VEGF receptors in cultured lymphatic endothelial cells. Microcirculation 6:75–78
10. Yokoyama S, Ohhashi T (1993) Effects of acetylcholine on spontaneous contractions in isolated bovine mesenteric lymphatics. Am J Physiol 264:H1460–H1464
11. Tunemoto H, Ikomi F, Ohhashi T (2003) Flow-mediated release of nitric oxide from lymphatic endothelial cells of pressurized thoracic duct. Jpn J Physiol 53:153–163
12. Shirasawa Y, Ikomi F, Ohhashi T (2000) Physiological roles of endogenous nitric oxide in lymphatic pump activity of rat mesentery in vivo. Am J Physiol 278:G551–G556
13. Jain RK (1990) Vascular and interstitial barrier to delivery of therapeutic agents in tumors. Cancer Metastasis Rev 9:253–266
14. Tan YH (1998) Basic fibroblast growth factor-mediated lymphangiogenesis of lymphatic endothelial cells isolated from dog thoracic ducts. Jpn J Physiol 48:133–141

How Does Inhaled Nitric Oxide Reach Peripheral Tissues?

Paul Kubes

Summary. Conventional wisdom would dictate that nitric oxide is a local autocoid with spatially limited effects. Over the last few years, we, and others have challenged this view and have used inhaled nitric oxide to demonstrate that despite its administration in lung, it can impact the peripheral vasculature. This chapter summarizes some of the evidence to support the contention that nitric oxide can impact peripheral vasculatures presumably via a stabilizing moiety in the circulation. One possibility is the formation of *S*-nitrosothiols, which extend the half-life of nitric oxide many-fold. In this chapter I provide evidence that *S*-nitrosothiols exist in the vasculature, particularly during nitric oxide inhalation. Finally, I highlight the limited evidence for the role that these potent vasodilating molecules may play as physiologically and therapeutically important regulators of the vascular system.

Key words. Nitrosothiol, *S*-Nitrosoalbumin, Ischemia/reperfusion, Microcirculation

Introduction

For many years the diatomic free radical nitric oxide was considered exclusively as an atmospheric pollutant produced during the combustion of fossil fuels [1,2]. It is now well appreciated that nitric oxide is produced by a large variety of organisms including all mammals [3–5]. Nitric oxide has many essential important functions including blood pressure regulation, host

Immunology Research Group, Department of Physiology and Biophysics, University of Calgary, Health Sciences Center, 3330 Hospital Drive NW, Calgary, AB, T2N 4N1, Canada defense, and neurotransmission [4,6]. However, conventional wisdom dictates that these effects are mediated locally, by NO-producing cells. In fact, NO is rapidly inactivated by oxyhemoglobin (HbO_2, i.e., $Hb[Fe_{II}]O_2$) to form methemoglobin (MetHb, i.e., $Hb[Fe_{III}]$) and nitrate (NO_3^-) within the bloodstream, thereby restricting its actions to the site of production. Herein, I describe the possibility that in fact nitric oxide may be stabilized in the circulation and may have much further-reaching effects than was previously proposed.

RSNOs, Vasodilators Formed in Blood

Loscalzo, Stamler, and other colleagues challenged the view that nitric oxide was restricted to local effects when they proposed that nitric oxide can potentially bind carrier molecules to form nitric oxide adducts called *S*-nitrosothiols (RSNOs). These molecules were shown to function as nitric oxide-carrying systems, prolonging the half-life and spatial impact of nitric oxide [7,8]. First, plasma RSNOs have been detected in vivo by many investigators [7–13]. Examples of RSNOs include low molecular weight *S*-nitrosocysteine (CysNO), *S*-nitrosoglutathione (GSNO), and high molecular weight *S*-nitroso-albumin (SNO-Alb). Among these RSNOs, SNO-Alb tends to be more stable than low molecular weight molecules [7] and is the principal molecule formed [7]. Injection of any of these RSNOs into animals results in prolonged vasodilation [7,8,14,15]. Therefore, RSNOs can be formed in vivo and when synthesized ex vivo and injected into animals, they have vasodilating properties. The question that remained was whether RSNO can be produced in sufficient quantities in vivo to function as circulating vasodilators.

Inhaled Nitric Oxide Affects Peripheral Vascular Beds

Nitric oxide is administered directly by inhalation for the treatment of pulmonary hypertension in newborn infants [16,17] and acute respiratory distress syndrome in adults [4,16]. Although this form of nitric oxide delivery is used for regional impact exclusively on the pulmonary vasculature, we made use of this system to ask whether the nitric oxide administered in the lung could exert its biology beyond the lung. This would indeed support the production of NO carriers in vivo. However, our initial attempts failed miserably. Inhalation of 80 ppm nitric oxide did not in any way affect blood pressure or, for example, intestinal blood flow. This was consistent with similar reports by others that inhalation of nitric oxide does not affect basal physiology. However, when nitric oxide was systemically inhibited and systemic blood pressure was increased, inhaled nitric oxide reduced but did not prevent the rise in blood pressure. However, the problem with this experiment was that the systemic changes in blood pressure could simply reflect changes within the pulmonary microvasculature. Therefore, Fox-Robichaud and colleagues next decided to inhibit nitric oxide production locally within a small distal microvasculature [18]. This was accomplished by superfusing the mesenteric microvasculature and using intravital microscopy to visualize changes within the microvessels of the mesentery. When nitric oxide was inhibited, profound increase in vasoconstriction and leukocyte adhesion was noted in the mesenteric microvessels. When animals were made to breath inhaled nitric oxide, the vasoconstriction was no longer detectable and leukocyte adhesion was greatly reduced. Clearly, delivery of nitric oxide at the lung, somehow affected the mesenteric microvessels.

One potential criticism of this work was that at least the leukocyte effects could potentially have occurred within the lung making the leukocytes no longer adhere in the periphery. To address whether the leukocytes were being affected within the lung, the animals were made to breathe nitric oxide, and their blood was immediately taken and perfused through a flow chamber containing adhesion molecules (surrogate blood vessel). The data from this series of experiments clearly revealed that leukocytes adhered as effectively when taken from animals breathing room air or nitric oxide [18]. Clearly the nitric oxide was not affecting the leukocytes, but rather was reaching the distal microvasculature and modulating the microvessels. Indeed, this conclusion was also consistent with the inhibition of vasoconstriction in the peripheral blood vessels with inhaled nitric oxide wherein leukocytes were presumably not involved. Moreover, the vasoconstriction data were reproduced in humans. Responses to inhibition of local NO synthesis was reversed following administration of inhaled NO [18,19].

Inhaled Nitric Oxide Affects Pathophysiology in Peripheral Organs

Although the reversal of nitric oxide inhibition by inhaled nitric oxide was important support that nitric oxide could reach peripheral vasculatures, from a therapeutic standpoint, it remained unclear whether this approach could improve pathology. Fox-Robichaud and colleagues tested this hypothesis and demonstrated that inhaled NO was also beneficial in pathological conditions wherein NO has been documented to be reduced. Inhaled NO reduced leukocyte recruitment and prevented reductions in blood flow caused by ischemia/reperfusion [18,20]. Interestingly, when the mesentery was exposed to lipopolysaccharide which induced profound leukocyte recruitment, inhaled nitric oxide was not able to reduce this adhesion [18]. The difference between the two models was that in ischemia/reperfusion nitric oxide is low or completely inhibited whereas in the endotoxemia model increased NO production is a key feature. Clearly, inhaled nitric oxide will not be beneficial in all vascular diseases.

The next important question was whether there was an increase in RSNOs during NO inhalation. Although both Cannon and colleagues and Kubes and colleagues observed an increase in NO-carrying molecules, the specific molecules detected were quite different. Cannon and colleagues [19] demonstrated that HbNO in blood rose significantly during NO breathing, they were not able to detect a change in SNO-Alb in plasma with NO inhalation. This is in contrast to our own data wherein we did observe an increase in SNO-Alb (Ng and Kubes (2004) Circ Res 94(4):559–565). A very likely explanation for this difference, which in fact was raised by Cannon and colleagues, is that SNO-Alb decomposes rapidly due to the interaction of this species with low molecular weight thiol groups [19]. Addition of sulfanilamide, diethylenetriaminepentaacetic acid (DTPA), and *N*-ethylmaleimide during sampling can reduce SNO-Alb degradation and transnitrosation reactions.

Other investigators have delivered nitric oxide directly into the bloodstream (not through inhalation) and also observed RSNO formation. Marley and colleagues [21] studied the formation of RSNOs from low fluxes of NO in plasma. They showed that significant amounts of RSNOs and more specifically, SNO-Alb were formed. Rassaf and colleagues [22] also demonstrated the formation of plasma RSNOs in vivo following infusion of NO into the bloodstream. Again more than 90% of the high molecular weight RSNOs were SNO-Alb. Rassaf et al. [22] observed both a rapid vasodilation consistent with administration of exogenous NO and a delayed vasodilation temporally similar to effects observed with the administration of RSNOs.

Another important question was whether inhaled NO was restricted to the microvasculature or whether it was also able to impact on sites outside the vasculature. Inhibition of nitric oxide in the intestine caused an increase in epithelial or mucosal permeability [20]. However, inhaled nitric oxide was unable to reduce the increased mucosal permeability whereas NO donors did have biological activity at these extravascular sites. These data suggest that the NO delivery system was restricted to the vasculature. When lymphatics were cannulated and the RSNOs measured, there was absolutely no notable increase in these molecules. Clearly, these observations suggest that RSNOs can form in the vasculature but they are unable to reach the extravascular space. One possibility is that the nitric oxide is removed from the RSNO by endothelium or that the main RSNO is too large to reach the extravascular space. One example of this would be SNO-hemoglobin inside red blood cells, which do not enter the extravascular space.

How Might RSNOs Be Formed In Vivo?

As a free radical, NO is highly reactive and short-lived with a half-life of only 0.05–1 s in blood [7,23,24]. Although NO is thought not to react directly with thiols, a number of indirect reactions for RSNO formation have been proposed. In the aqueous phase of plasma, NO may react with O_2 to form higher oxides of nitrogen (N_2O_3), which subsequently leads to the formation of nitrite (NO_2^-). On the other hand, the intermediate, N_2O_3 (a carrier of NO^+), formed from the autoxidation of NO, may undergo *S*-nitrosylation with molecules containing thiol groups to form RSNOs [8,25,26]. Although a number of these reactions have been demonstrated in vitro, the exact mechanisms by which RSNOs are formed in vivo remain uncertain due to the complexity of blood leading to numerous other competitive reactions in biological systems [21,25,27].

Nitric oxide interacts with superoxide (O_2^-) to form peroxynitrite ($ONOO^-$) in plasma. $ONOO^-$ decomposes rapidly once protonated to generate either nitrate (NO_3^-) [23,24,28] or strong oxidants such as hydroxyl and nitrogen dioxide radicals. Reactions between $ONOO^-$ and thiols have been reported leading to the formation of RSNOs [29–33]. The observation that administration of exogenous $ONOO^-$ into animals induced responses reminiscent of RSNOs including vasodilation [29,33] and inhibition of leukocyte–endothelial cell interactions [34] potentially supports the view that $ONOO^-$ rapidly reacts with thiols to form RSNOs.

Finally, metabolic pathways of NO in erythrocytes other than to form nitrate via oxyhemoglobin may function as important NO delivery systems. For example, a small proportion of NO may bind to deoxyhemoglobin (Hb, i.e., Hb[Fe_{II}]) to form nitrosylhemoglobin (HbNO, i.e., Hb[Fe_{II}]NO), or with the 93-cysteine residue of the β-subunit to form *S*-nitrosohemoglobin (SNO-Hb) [24,35–37]. The fraction of each is dependent on the ratio of oxygenated and deoxygenated Hb within the erythrocytes [23,24]. These molecules have been detected in vivo and can induce vasodilation when formed exogenously in red blood cells and reinjected into animals. The question that remains is whether sufficient amounts of these hemoglobin molecules are produced in vivo to affect peripheral vasculatures.

Concluding Remarks

There is a growing body of evidence that nitric oxide is not a local autocoid that influences only its immediate environment. It is now becoming apparent that nitric oxide may combine with thiols and be transported around the circulatory system, perhaps delivering nitric oxide where it is needed most.

Inhaled nitric oxide is a simple and effective tool to begin to demonstrate that nitric oxide can indeed bind molecules within the blood to impact extrapulmonary vasculatures. Since inhaled nitric oxide is a reasonably safe method for nitric oxide delivery, it could potentially be important as a therapeutic modality in various cardiovascular diseases.

References

1. Burgner D, Rockett K, Kwiatkowski D (1999) Nitric oxide and infectious diseases. Arch Dis Child 81:185–188
2. Keh D, Gerlach H, Falke K (2000) Inhalation therapy with nitric oxide gas. In: Mayer B (ed) Nitric oxide. Springer, Berlin Heidelberg New York, pp 399–441
3. Ignarro LJ, Buga GM, Wood KS, et al (1987) Endothelium-derived relaxing factor produced and released from artery and vein is nitric oxide. Proc Natl Acad Sci USA 84:9265–9269
4. Moncada S, Higgs A (1993) The L-arginine-nitric oxide pathway. N Engl J Med 329: 2002–2012
5. Palmer RMJ, Ferrige AG, Moncada S (1987) Nitric oxide release accounts for the biological activity of endothelium-derived relaxing factor. Nature 327:524–526
6. Moncada S, Palmer RMJ, Higgs EA (1991) Nitric oxide: physiological, pathophysiology, and pharmacology. Pharmacol Rev 43:109–142
7. Stamler JS, Jaraki O, Osborne J, et al (1992) Nitric oxide circulates in mammalian plasma primarily as an S-nitroso adduct of serum albumin. Proc Natl Acad Sci USA 89:7674–7677
8. Stamler JS, Simon DI, Osborne JA, et al (1992) S-nitrosylation of proteins with nitric oxide: synthesis and characterization of biologically active compounds. Proc Natl Acad Sci USA 89:444–448
9. Goldman RK, Vlessis AA, Trunkey D (1998) Nitrosothiol quantification in human plasma. Anal Biochem 259:98–103
10. MacMicking J, Xie QW, Nathan C (1997) Nitric oxide and macrophage function. Annu Rev Immunol 15:323–350
11. Marzinzig M, Nussler AK, Stadler J, et al (1997) Improved methods to measure end products of nitric oxide in biological fluids: nitrite, nitrate, and S-nitrosothiols. Nitric Oxide Biol Chem 1:177–189
12. Minamiyama Y, Takemura S, Inoue M (1997) Effect of thiol status on nitric oxide metabolism in the circulation. Arch Biochem Biophys 341:186–192
13. Scharfstein JS, Keaney JF, Slivka A, et al (1994) In vivo transfer of nitric oxide between a plasma protein-bound reservoir and low molecular weight thiols. J Clin Invest 94: 1432–1439
14. Myers PR, Minor RL Jr, Guerra R Jr, et al (1990) Vasorelaxant properties of the endothelium-derived relaxing factor more closely resemble S-nitrosocysteine than nitric oxide. Nature 345:161–163
15. Rubanyi GM, Johns A, Wilcox D, et al (1991) Evidence that a S-nitrosothiol, but not nitric oxide, may be identical with endothelium-derived relaxing factor. J Cardiovasc Pharmacol 17:S41–S45

16. Kinsella JP, Neish SR, Shaffer E, et al (1992) Low-dose inhalational nitric oxide in persistent pulmonary hypertension of the newborn. Lancet 340:819–820
17. Roberts JD, Polaner DM, Lang P, et al (1992) Inhaled nitric oxide in persistent pulmonary hypertension of the newborn. Lancet 340:818–819
18. Fox-Robichaud A, Payne D, Hasan SU, et al (1998) Inhaled NO as a viable antiadhesive therapy for ischemia/reperfusion injury of distal microvascular beds. J Clin Invest 101:2497–2505
19. Cannon RO III, Schechter AN, Panza JA, et al (2001) Effects of inhaled nitric oxide on regional blood flow are consistent with intravascular nitric oxide delivery. J Clin Invest 108:279–287
20. Kubes P, Payne D, Grisham MB, et al (1999) Inhaled NO impacts vascular but not extravascular compartments in postischemic peripheral organs. Am J Physiol (Heart Circ Physiol) 277:H676–H682
21. Marley R, Patel RP, Orie N, et al (2001) Formation of nanomolar concentrations of S-nitroso-albumin in human plasma by nitric oxide. Free Radic Biol Med 31:688–696
22. Rassaf T, Preik M, Kleinbongard P, et al (2002) Evidence for in vivo transport of bioactive nitric oxide in human plasma. J Clin Invest 109:1241–1248
23. Kelm M (1999) Nitric oxide metabolism and breakdown. Biochim Biophys Acta 1411: 273–289
24. Kelm M, Yoshida K (1996) Metabolic fate of nitric oxide and related N-oxides. In: Feelisch M, Stamler JS (eds) Methods in nitric oxide research. Wiley, West Sussex, pp 47–58
25. Kharitonov VG, Sundquist AR, Sharma VS (1995) Kinetics of nitrosation of thiols by nitric oxide in the presence of oxygen. J Biol Chem 270:28158–28164
26. Stamler JS, Singel D, Loscalzo J (1992) Biochemistry of nitric oxide and its redox-activated forms. Science 258:1898–1902
27. Nedospasov A, Rafikov R, Beda N, et al (2000) An autocatalytic mechanism of protein nitrosylation. Proc Natl Acad Sci USA 97:13543–13548
28. Beckman JS, Beckman TW, Chen J, et al (1990) Apparent hydroxyl radical production by peroxynitrite: implications for endothelial injury from nitric oxide and superoxide. Proc Natl Acad Sci USA 87:1620–1624
29. Liu S, Beckman JS, Ku DD (1994) Peroxynitrite, a product of superoxide and nitric oxide, produces coronary vasorelaxation in dogs. J Pharmacol Exp Ther 268:1114–1121
30. Moro MA, Darley-Usmar V, Goodwin DA, et al (1994) Paradoxical fate and biological action of peroxynitrite on human platelets. Proc Natl Acad Sci USA 91:6702–6706
31. Nossuli T, Hayward R, Scalia R, et al (1997) Peroxynitrite reduces myocardial infarct size and preserves coronary endothelium after ischemia and reperfusion in cats. Circulation 96:2317–2324
32. van der Vliet A, Chr't Hoen PA, Wong PSY, et al (1998) Formation of S-nitrosothiols via direct nucleophilic nitrosation of thiols by peroxynitrite with elimination of hydrogen peroxide. J Biol Chem 273:30255–30262
33. Wu M, Pritchard KA Jr, Kaminski PM, et al (1994) Involvement of nitric oxide and nitrosothiols in relaxation of pulmonary arteries to peroxynitrite. Am J Physiol (Heart Circ Physiol) 266:H2108–H2113
34. Lefer DJ, Scalia R, Campbell B, et al (1997) Peroxynitrite inhibits leukocyte-endothelial cell interactions and protects against ischemia-reperfusion injury in rats. J Clin Invest 99:684–691

35. Gow AJ, Stamler JS (1998) Reactions between nitric oxide and hemoglobin under physiological conditions. Nature 391:169–173
36. Jia L, Bonaventura C, Bonaventura J, et al (1996) S-nitrosohemoglobin: a dynamic activity of blood involved in vascular control. Nature 380:221–226
37. Stamler JS, Jia L, Eu JP, et al (1997) Blood flow regulation by S-nitrosohemoglobin in the physiological oxygen gradient. Science 276:2034–2037

Tumor Angiogenesis Regulated by Gaseous Molecules in Tumor Microenvironment: Oxygen, pH, and Nitric Oxide

Dai Fukumura

Summary. Hypoxia and acidosis are hallmarks of metabolic environment in solid tumors and regulate expression of vascular endothelial growth factor (VEGF), a key angiogenesis factor. We developed a novel in vivo microscopy technique to simultaneously measure VEGF promoter activity, pO_2 and pH. To monitor VEGF expression in vivo, we engineered human glioma cells that express green fluorescent protein (GFP) under the control of VEGF promoter. Tissue pO_2 and pH were determined by phosphorescence quenching microscopy and ratio imaging microscopy, respectively. These techniques have allowed us to show that VEGF transcription in brain tumors is regulated by tissue pO_2 and pH via distinct pathways. Nitric oxide (NO) is a multi functional gaseous molecule and regulates various physiological functions. We have shown that NO mediates vessel tone, blood flow, vascular permeability and leukocyte-endothelial interactions in solid tumors. Furthermore, we found that endothelial NO synthase (eNOS) plays predominant role in VEGF-induced angiogenesis and vascular permeability using angiogenic gel model in NOS deficient mice. These findings and the resulting mechanistic insights in the regulation of angiogenesis by gaseous molecules have significant implications for novel tumor detection and treatment strategies.

Key words. pO_2, pH, Vascular endothelial growth factor, Tumor, Nitric oxide

Introduction

Tumor vessels have tortuous shape, irregular surface and diameter, and heterogeneous spatial distribution [1]. Tumor blood flow is often sluggish and static, and even changes the direction over time. Resulting hypoxia and

---

Department of Radiation Oncology, Edwin L. Steele Laboratory, Massachusetts General Hospital, Harvard Medical School, 100 Blossom Street, COX-7, Boston, MA 02114, USA acidosis become hallmarks of tumors and critical determinants of tumor growth, metabolism, and response to conventional and novel treatments. Vascular endothelial growth factor (VEGF) is one of the most potent angiogenic factors, expressed in a wide variety of tumors, and correlated with angiogenesis, tumor growth, invasion, metastasis and prognosis [2]. Hypoxia upregulates VEGF via hypoxia inducible factor (HIF) in various types of cells including tumor cells [2]. However, we and others found a lack of spatial correlation between hypoxia and VEGF expression in solid tumors [3,4]. Subsequently, we found that acidic pH, another characteristic tumor metabolic environment regulates VEGF expression via distinct pathway from hypoxia and thus, provides complementary regulation of VEGF in solid tumors [5,6].

Nitric oxide (NO) regulates various physiological functions including angiogenesis [7]. Nitric oxide not only induces endothelial migration and proliferation by itself but also mediates effects of many of angiogenic factors such as VEGF. We found that endothelial NO synthase (eNOS) plays predominant role in VEGF-induced angiogenesis and vascular permeability [8]. These findings and the resulting mechanistic insights in the regulation of angiogenesis by gaseous molecules have significant implications for novel tumor treatment strategies.

Vascular Endothelial Growth Factor is Regulated by pO_2 and pH Levels via Distinct Pathways

A prerequisite to measure the spatial correlation between tissue pO_2, pH, and VEGF promoter activity in vivo is the establishment of a technique which allows us to monitor the VEGF promoter activity without harvesting tissue. To accomplish this, we have engineered a red-shifted GFP vector driven by the VEGF promoter (*VEGF^P-GFP*) [9] and introduced it into the U87 (human glioblastoma) tumor cell line by electroporation. To test the system, we have determined the effect of hypoxia on the transfected cells in vitro. *VEGF^P-GFP* U87 cells were cultured under normoxia or hypoxia (1% oxygen). Green fluorescent protein fluorescence intensity was determined by FACS analysis. GFP fluorescence was significantly increased by hypoxia [5]. We also determined endogenous VEGF mRNA expression and GFP mRNA under this construct by Northern blot analysis. Both endogenous VEGF and GFP were upregulated by hypoxia in comparable ratio [5].

We implanted *VEGF^P-GFP* U87 tumors in SCID mouse cranial windows [1]. Seven to eight days after the implantation, tumors had grown and were well vascularized. GFP fluorescence in U87 tumor cells was visualized by fluorescence microscopy and translated GFP fluorescence intensity into instantaneous GFP concentration (Fig. 1) [9,10]. We used this setup with known

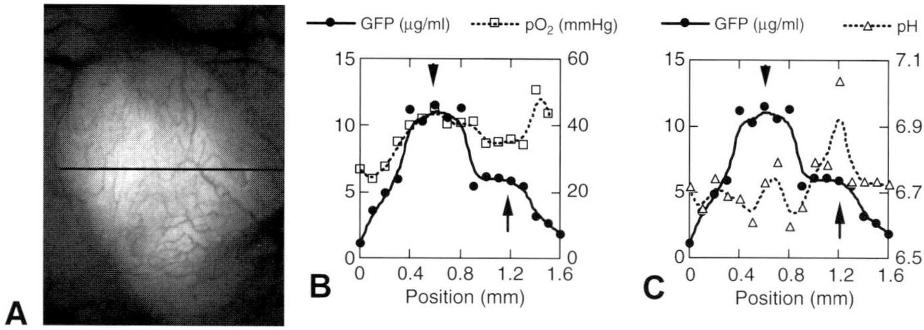

FIG. 1A–D. Simultaneous measurements of green fluorescent protein (*GFP*), pO_2, and pH. U87 tumor transfected with GFP gene driven by the promoter of vascular endothelial growth factor (VEGF) was scanned along the line indicated in **A**. VEGF promoter activity (**B,C**), tissue pO_2 (**B**), and pH (**C**) in the same spots were determined. Reproduced from [5]

quantities of recombinant EGFP to generate the necessary calibration curves. Tissue pO_2 and pH were determined by phosphorescence quenching microscopy and ratio imaging microscopy, respectively. Fig. 1 shows representative spatial correlation between VEGF promoter activity, tissue pO_2, and pH. In this tumor, the region that exhibits strongest GFP signal (*arrowhead*) was relatively well oxygenated (Fig. 1B), but showed low extracellular pH (<6.7) (Fig. 1C). The adjacent region showed a relative decrease in GFP (*arrow*). However, pO_2 did not significantly change (Fig. 1B), while tissue pH dramatically increases (Fig. 1C).

We analyzed the relationship between pO_2, pH, and VEGF expression in *VEGF^P-GFP* U87 tumor by dividing the measurements into either hypoxic ($pO_2 < 30$ mmHg) or oxygenated ($pO_2 > 30$ mmHg) groups and either low pH (pH < 6.8) or neutral (6.8 < pH < 7.4) groups. This allowed comparison of GFP expression under each of these conditions using linear regression. Figures 2A and D show that under hypoxic conditions or neutral conditions, pO_2, but not pH, is correlated to GFP expression. Conversely, Figures 2B and C show that under low pH conditions or oxygenated conditions, pH and not pO_2 is related to GFP expression. These results indicated for the first time that VEGF transcription in brain tumors is independently regulated by the tissue pO_2 and pH [5].

These findings prompted us to study signaling pathway to mediate acidic pH induced VEGF upregulation. Serial deletion of the VEGF promoter showed that the promoter region containing the AP-1 binding site but not hypoxia responsible element (HRE) or AP-2 and SP-1 binding sites was crucial for acidic-pH-induced VEGF upregulation. This is clearly different from hypoxia induced VEGF upregulation which is mediated by HIF-1 complex binding to

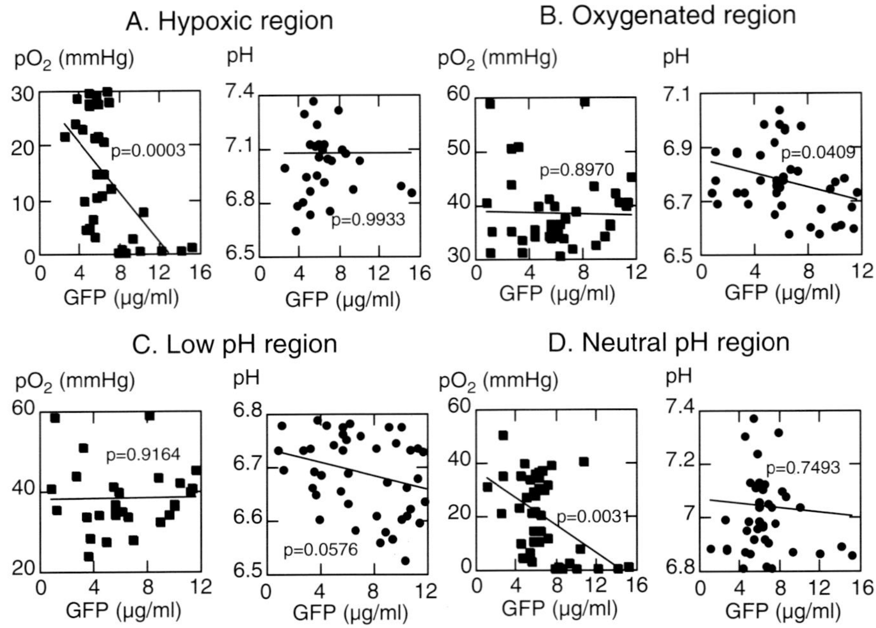

FIG. 2A–D. Relationships between VEGF promoter activity, tissue pO_2, and extracellular pH. *GFP*, green fluorescent protein. Reproduced from [5]

HRE. Acidic pH also stabilized VEGF mRNA. Furthermore, the acidic-pH-activates Ras and down stream signaling requires ERK1/2 MAPK but not p38 or JNK [6]. These data confirmed that two major metabolic environment in solid tumors complementarily regulate VEGF expression via distinct signaling pathways.

Endothelial NO Synthase Mediates Angiogenesis and Vascular Permeability

Nitric oxide is a gaseous molecule with an astonishingly wide physiological and pathophysiological activities. Nitric oxide is produced by a heme-containing enzyme, NOS, which is linked to nicotinamide adenine dinucleotide phosphate-derived electron transport via flavin adenine dinucleotide and flavin mononucleotide to catalyze oxidation of L-arginine to L-citrulline and NO with tetrahydrobiopterin as an essential cofactor. There are three isoforms of NOS. Vascular endothelial cells constitutively express endothelial NOS (eNOS, also referred to as type III NOS). Endothelial NOS activity is calcium-dependent and generates relatively low levels of NO. Nitric oxide produced by eNOS mediates a variety of physiological functions in vivo. In contrast, inducible NOS (iNOS, type II NOS) is calcium independent, transcriptionally regulated by inflammatory cytokines and other stimuli, and generates higher levels of NO. High amounts of NO can induce cytostatic/toxic effects. Finally, neuronal NOS (nNOS, type I NOS) mediates transmission of neuronal signals.

Nitric oxide mediates VEGF-induced angiogenesis and vascular hyperpermeability. However, the relative contribution of different NOS isoforms to these processes was not known. We evaluated the relative contributions of endothelial and inducible NOS (eNOS and iNOS, respectively) to angiogenesis and permeability of VEGF-induced angiogenic vessels [8]. The contribution of eNOS was assessed using an *eNOS*-deficient mouse and iNOS contribution was assessed using a selective inhibitor (L-NIL) and an *iNOS*-deficient mouse. Angiogenesis was induced by VEGF in type I collagen gels placed in the mouse cranial window. Angiogenesis in control collagen gels (Fig. 3A) and VEGF (3 μg/ml) containing gels (Fig. 3B) in mouse cranial window was monitored for 14 days. In the control gels, *eNOS*$^{-/-}$ mice showed significantly reduced angiogenesis compared to wild-type (WT) mice. L-NIL (an iNOS selective inhibitor) treatment showed slightly slower angiogenesis. In the VEGF gel, L-NIL treatment modestly reduced angiogenesis. Angiogenesis in *eNOS*$^{-/-}$ mice was significantly reduced compared to WT mice. L-NIL treatment of *eNOS*$^{-/-}$ mice further delayed angiogenesis. The rate of angiogenesis (Fig. 3C) showed that VEGF significantly accelerated angiogenesis in WT mice with or without L-NIL treatment, but not in *eNOS*$^{-/-}$ mice. In

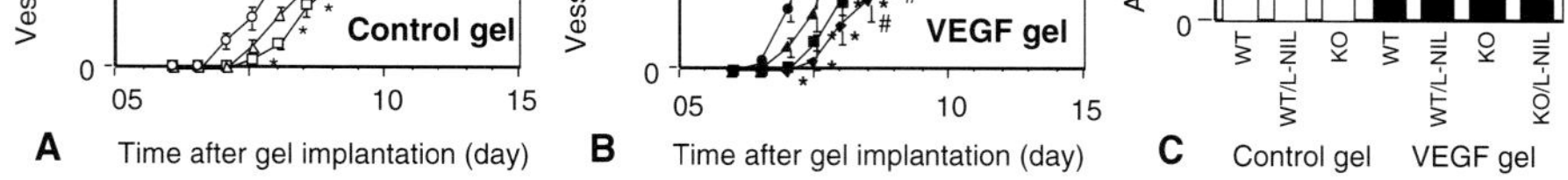

FIG. 3A–C. Role of nitric oxide synthase isoforms in vascular endothelial growth factor (*VEGF*)-induced angiogenesis. **A** Angiogenesis assay in control collagen gels. *Open squares*, *eNOS*$^{-/-}$ (*KO*) mice; *open circles*, wild-type (*WT*) mice; *open triangles*, WT with L-NIL treatment. **B** Angiogenesis assay in VEGF containing gels. *Closed squares*, *eNOS*$^{-/-}$ (*KO*) mice; *closed circles*, wild-type (*WT*) mice; *closed triangles*, WT with L-NIL treatment; *diamonds*, *eNOS*$^{-/-}$ mice with L-NIL. *eNOS*$^{-/-}$ mice (*KO/L-NIL*) further delayed angiogenesis. **C** Rate of angiogenesis. *$P < 0.05$ as compared with WT mice. $^{\#}P < 0.05$ as compared with WT mice with L-NIL treatment. Reproduced from [8]

both control gel and VEGF gel, L-NIL treatment slightly slowed the angiogenesis and KO showed significantly slower angiogenesis than WT mice. Inhibition of both eNOS and iNOS (KO/L-NIL) resulted in further delays in angiogenesis.

The role of NOS in VEGF-induced vascular permeability increase in vessels was also determined using *eNOS* and *iNOS*-deficient mice. For quiescent vessels, VEGF superfusion significantly increased permeability in both WT and *iNOS*$^{-/-}$ mice but not in *eNOS*$^{-/-}$ mice. Vascular permeability of newly formed vessels in VEGF gels in WT mice was significantly higher compared to the control gels in WT mice. eNOS but not iNOS inhibition in the VEGF gels resulted in significantly decreased permeability. These findings suggest that eNOS plays a predominant role in VEGF-induced angiogenesis and vascular permeability [8].

Recent data revealed the predominant role of eNOS in not only angiogenesis—development of new blood vessels derived from existing vessels—but also vasculogenesis—de novo blood vessel formation from stem/progenitor cells [11]. Aicher et al. [11] reported that *eNOS*$^{-/-}$ mice have a profoundly reduced basal activity of pro-matrix metalloproteinase (MMP)-9 and that 5-fluorouracil-induced MMP-9 is greatly decreased in the absence of eNOS. Consequently, membrane-bound Kit ligand cleavage is reduced and mice have decreased sKitL levels in the plasma. Functionally, *eNOS*$^{-/-}$ mice have impaired mobilization of the bone marrow progenitor cells, and wild-type cell bone marrow transplantation could not restore function of endothelial progenitor cells in an ischemic limb model. These data demonstrate that in the bone marrow, NO release activates MMP-9 as a critical paracrine interaction event for cell mobilization in circulation. Direct infusion of wild-type progenitor cells, but not of *eNOS*$^{-/-}$ cells, was able to restore defective neovascularization in *eNOS*$^{-/-}$ mice indicating that eNOS is also required at the site of vessel formation.

These data highlight again the roles of eNOS in neovascularization and thus, make selective modulation of eNOS activity an attractive strategy for altering angiogenesis and/or vascular permeability [12]. With the recent developments of an eNOS specific inhibitor and successful gene therapy approaches, modulation of eNOS may be a potent new strategy to control neovascularization for diseases with abnormal vessel formation. In fact, the first proof of this concept was recently published by Sessa and colleagues, who showed that selective inhibition of eNOS by cavtratin blocks tumor vessel hyperpermeability and halts tumor progression [13].

Several key questions remain, however, unanswered on the role of gaseous molecules on vasculature. Elucidating the role of these molecules on vessel specification/maturation would be of great interest for tissue engineering and drug delivery. The downstream signaling of NO is also important to design alternative strategies to control NO-dependent effects on neovascularization. For instance, cyclic guanosine monophosphate (cGMP) is known as the second messenger of vasodilation induced by NO. Intracellular concentration of cGMP is tightly regulated by phosphodiesterases. Since phosphodiesterase inhibitors are now in clinical use, it would be critical to document the effects of phosphodiesterase inhibitors on neovascularization. The role of carbon monoxide, another putative regulator of physiological functions on angiogenesis and tumor microcirculation is still largely understudied. Finally, lymphatic vessel function is equally important in maintaining tissue homeostasis, and disruption of the lymphatic system is encountered in many pathologies. Compared to the blood vessels, our knowledge of oxygen, CO and NO functions on lymphatic vessels is in its infancy. A better understanding of the gaseous molecule biology will expand the horizon of its physiological and pathological functions and lead us to new strategies to treat various diseases.

Acknowledgments. This work was supported by NIH Grants PO1-CA80124 (to D.F. and Rakesh K. Jain) and RO1-CA96915 (to D.F.).

References

1. Jain RK, Munn LL, Fukumura D (2002) Dissecting tumor pathophysiology using intravital microscopy. Na Rev Cancer 2:266–276
2. Carmeliet P, Jain RK (2000) Angiogenesis in cancer and other diseases: from genes to function to therapy. Nature 407:249–257
3. Fukumura D, Chen Y, Gohongi T, et al (1998) VEGF expression in tumors is regulated by oxygen gradient rather than absolute level of hypoxia itself. Proc Am Assoc Cancer Res 39:271
4. Raleigh JA, Calkins-Adams DP, Rinker LH, et al (1998) Hypoxia and vascular endothelial growth factor expression in human squamous cell carcinoma using pimonidazole as a hypoxia marker. Cancer Res 58:3765–3768
5. Fukumura D, Xu L, Chen Y, et al (2001) Hypoxia and acidosis independently up-regulate vascular endothelial growth factor transcription in brain tumors in vivo. Cancer Res 61:6020–6024
6. Xu L, Fukumura D, Jain RK (2002) Acidic extracellular pH induces vascular endothelial growth factor (VEGF) in human glioblastoma cells via ERK1/2 MAPK signaling pathway. Mechanism of low pH induced VEGF. J Biol Chem 277:11368–11374
7. Fukumura D, Jain RK (1998) Role of nitric oxide in angiogenesis and microcirculation in tumors. Cancer Metastas Rev 17:77–89
8. Fukumura D, Gohongi T, Kadambi A, et al (2001) Predominant role of endothelial nitric oxide synthase in vascular endothelial growth factor-induced angiogenesis and vascular permeability. Proc Natl Acad Sci USA 98:2604–2609
9. Fukumura D, Xavier R, Sugiura T, et al (1998) Tumor induction of VEGF promoter in stromal cells. Cell 94:715–725

10. Brown EB, Campbell RB, Tsuzuki Y, et al (2001) In vivo measurement of gene expression, angiogenesis, and physiological function in tumors using multiphoton laser scanning microscopy. Nat Med 7:864–868
11. Aicher A, Heeschen C, Mildner-Rihm C, et al (2003) Essential role of endothelial nitric oxide synthase for mobilization of stem and progenitor cells. Nat Med 9:1370–1376
12. Duda DG, Fukumura D, Jain RK (2004) Role of eNOS in neovascularization: NO for endothelial progenitor cells. Trends Mol Med 10:143–145
13. Gratton JP, Lin MI, Yu J, et al (2003) Selective inhibition of tumor microvascular permeability by cavtratin blocks tumor progression in mice. Cancer Cell 4:31–39

Oxygen-Induced Cytoskeleton Rearrangement of Cultured Human Brain Microvascular Endothelial Cells

Minoru Tomita[1], Norio Tanahashi[1], Masahiro Kobari[2], Hidetaka Takeda[1], Masaki Takao[1], and Istvan Schiszler[1,3]

Key words. Oxygen, Human brain microvascular endothelial cell, Remodeling of cytoskeleton, Sol-gel transformation, Mobile cell

Introduction

Dynamic remodeling of endothelial cells in response to various stimuli is important for capillary flow redistribution. We previously reported that exposure of nonconfluent cultured human brain microvascular endothelial cells (HBEC) to oxygen induced a contraction by 14%, with formation of multiplex mesh networks consisting of star shapes, large and small polygons, and fern and brush patterns [1]. Cessation of oxygen supply induced disassembly of the network, incorporation of decomposed particles into cell components (or solation into the fluid), and recovery of the previous cell complexion. This paper examines the reproducibility of these sequential processes, i.e., contraction–network formation in the cell–disintegration of fibers–particulation–recovery of the original cell shape, upon repeated oxygen exposure of the same HBECs.

Materials and Methods

The materials and methods used here were the same as those reported previously [1]. Briefly, we used nonconfluent HBECs of passage 2 purchased from Cell Systems Corp. (WA, USA). During observation with video-enhanced

[1] Department of Neurology, Keio University School of Medicine, 35 Shinanomachi, Shinjuku-ku, Tokyo 160-8582, Japan
[2] Kyosai Tachikawa Hospital, 4-2-22 Nishiki-cho, Tachikawa, Tokyo 190-0022, Japan
[3] (Present address) Petofi utca 4, Torokbalint 2045, Hungary

contrast differential interference contrast (VEC) microscopy, oxygen was directly introduced repeatedly (more than three times; maximum eight times) onto the medium surface above HBEC spread on a coverslip ($n = 6$) for periods of approximately 30 s at intervals of approximately 30 s.

Results

Human brain microvascular endothelial cells contained a round-topped nucleus with one or two nucleoli, and a cell body surrounded by a thin, widely spread transparent lamella, as described previously. We also confirmed that transient oxygen application produced the sequential processes of contraction–network formation in the cell–disintegration of fibers–particulation–recovery of the original cell shape. The peripheral lamella was dragged centrally towards the nucleus and the nuclear envelope became clearly demarcated and enhanced. The plasma membrane became tightly wrapped around the cell body, which then also began to contract. When the contraction slowed down and stopped, a mesh network suddenly appeared, starting at adhesion plaques, and grew rapidly like a spider's web. Such networks spread from the adhesion plaques of the HBEC to the lamella, as well as to the cell body and the surface of the nucleus. We found that repeated oxygen delivery to the same cells reproduced the sequential processes described above every time when oxygen was delivered. Figure 1 shows a typical protocol of three reproducible cycles in a HBEC. It should be noted that the patterns of the mesh network were different each time, i.e., large polygonal pattern at the first exposure, fern pattern at the second exposure, and small polygonal pattern at the third exposure. After discontinuance of the oxygen blowing each time, a recovery phase followed: the network disintegrated immediately, yielding small particles of $<0.5\,\mu m$ in diameter which subsequently fused into the cellular structure. The HBEC completely recovered the control appearance each time. The other cells studied showed similar reproducibility. Time intervals (average of six cases) were approximately 8.8 ± 4.3 s for contraction, 2.1 ± 3.1 s for network appearance, 14.4 ± 9.1 s for network duration, and 18.8 ± 13.6 s for disappearance. The overall process was therefore rapid and, as noted above, could be reproduced in individual cells each time that oxygen gas was supplied. Thus, repeated oxygen stimuli remodeled the cytoskeleton of HBECs. The changes in the mesh network (cytoskeleton) in the same HBECs were consistent in all six cells studied. There seemed to be no rules as to pattern formation, although the patterns were tentatively categorized into fern, polygonal, star-shaped, and brush-like types.

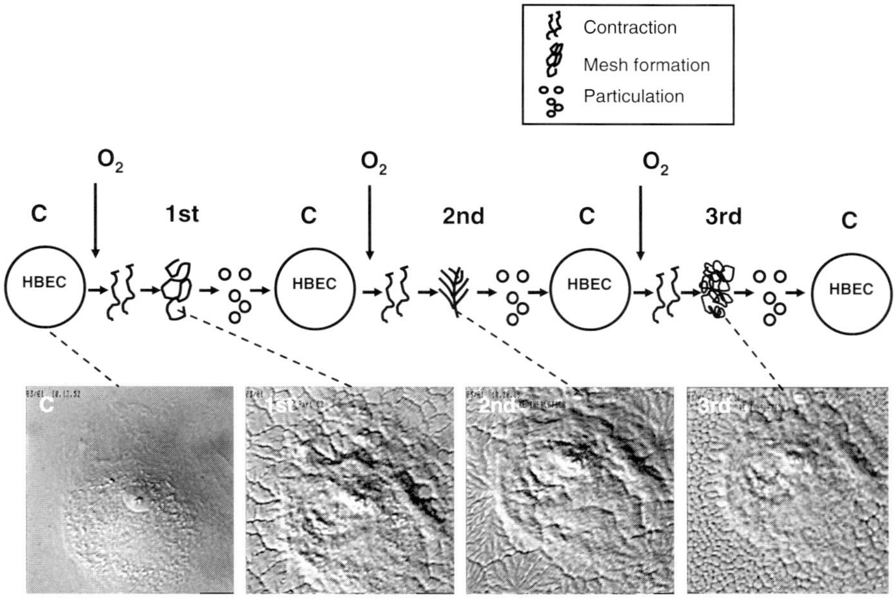

FIG. 1. Note that the patterns of mesh work were different each time of exposures to oxygen gas (see text for explanation). *C*, control; *1st*, 1st exposure to oxygen gas; *HBEC*, human brain microvascular endothelial cell

Discussion

The network involved could consist of stress fibers, which are believed to contain all the elements required for active contraction; myosin, ·-actin, and tropomyosin. It is likely that brain microvascular endothelial cells are remodeling almost continuously in response to various changing stimuli. Thus, we consider that capillaries do not behave as rigid tubes, or as compliant vessels. The remodeling upon contraction creates a capillary vasomotion not only radially, but also in a longitudinal direction, that may control the redistribution of capillary flow, along with local function. The above sequential changes with development of stress fibers and adhesion plaques indicate that HBEC has many of the characteristics of mobile cells.

Conclusion

We conclude that the cytoskeleton of HBECs rapidly and reproducibly remodels after successive contraction cycles in response to multiple oxygen exposures. The appearance of a mesh network and adhesion plaques suggests that HBECs are mobile cells.

Reference

1. Inoue K, Tomita M, Fukuuchi Y, et al (2003) Dynamic observation of oxygenation-induced contraction of and transient fiber-network formation/disassembly in cultured human brain microvascular endothelial cells. J Cereb Blood Flow Metab 23(7):821–828

Key Word Index